Automation Engineering

Yuri A.W. Shardt

Automation Engineering

Yuri A. W. Shardt
Technical University of Ilmenau
Erfurt, Germany

ISBN 978-3-031-92393-7   ISBN 978-3-031-92394-4 (eBook)
https://doi.org/10.1007/978-3-031-92394-4

This Springer imprint is published by the registered company Springer Nature Switzerland AG
The registered company address is: Gewerbestrasse 11, 6330 Cham, Switzerland

Preface

This book examines the foundations of automation engineering in a world increasingly focused on the development and implementation of automation. It provides the reader with an understanding of the key principles and components of automation engineering and how these principles can then be combined and implemented in a real industrial system to provide safe, economically viable, and efficient systems. A wide range of different systems, including chemical, electrical, and mechanical systems, will be considered.

At the end of each chapter, questions testing the reader's understanding of the material presented as well as providing places for extension and deeper understanding of the topics at hand are given.

In order to clearly differentiate between regular text and computer-related symbols, the font `Courier New` is used for all computer-related symbols.

This book is meant to be used as the course material for an introductory, bachelor's-level course in automation engineering. It is possible to present different permutations and combinations of the material depending on what topics are of interest to the particular students.

Computer files and related material can be downloaded from the book website (https://link.springer.com/book/9783031923937).

The author would like to thank Ying Deng and M. P. for their help in preparing some of the material used in this book. As well, the author would like to thank the students of the course AT.215 Automation Engineering for their corrections and suggestions for the contents of this textbook.

Erfurt, Germany Yuri A. W. Shardt

Competing Interests The author has no competing interests to declare that are relevant to the content of this manuscript.

Contents

List of Figures

List of Tables

List of Examples

Chapter 1
Introduction to Automation Engineering

Automation engineering is an important component of modern industrial systems that focuses on the development, analysis, optimisation, and implementation of complex systems to provide safe, economically viable, and efficient processes. Automation engineering seeks to eliminate as much as possible human intervention into the process. However, it should be noted that this does not mean that humans are not required to monitor and assist with the running of the process; it simply implies that the mundane, often repetitive, tasks are delegated to computer systems that are better suited for performing such work.

In order to understand automation engineering, it is important to briefly review its long history, its main principles, and its foundations.

1.1 The History of Automation Engineering

Ever since humans developed the need to implement complex tasks, the desire for making them faster and easier was also present. All such endeavours have sought to harness the power of nature using ingenious methods to provide the desired outcomes. Often, these devices have had practical or military implications.

One of the first to develop an interest in automating processes were the ancient Greeks, who developed a wide range of devices. Since these devices acted on their own, they were often called *automata* (singular: automaton; from the Greek αὐτόματον,[1] which means *acting of one's own will*). These automata performed a wide variety of tasks. Homer (gr: Ὅμηρος, *c.* eighth century BC) was the first to describe such devices as automated moving temple doors or tripods. The first

[1] This word is ultimately derived from the Greek word αὐτός, meaning *self*, and an unattested root, which comes from the proto-Indo-European word **méntis ~ mn̥téis*, meaning *thought* (which in turn gave us such words as *mind* in English).

Y. A. W. Shardt, *Automation Engineering*, https://doi.org/10.1007/978-3-031-92394-4_1

known device with feedback control is the water clock developed by Ctesibius (gr: Κτησίβιος; fl. 285–222 BC) that was able to accurately maintain the time. In fact, this water clock remained the most accurate time-keeping device until the invention of the pendulum clock in AD 1656 by Christiaan Huygens. Later, a primitive steam engine, called an æolipile, was developed by Hero of Alexandria (gr: Ἥρων ὁ Ἀλεξανδρεύς, *c.* AD 10–70). Examples of these devices are shown in Fig. 1.1. As well, objects that assisted in the computation of heavenly bodies (essentially the first computers) were also developed. The most famous is the Antikythera mechanism, a type of astronomical clock based on a gear-driven apparatus. This tradition of developing mechanical automated objects was continued long into the Middle Ages both in Europe and in the Middle East, with such work as the *Book of Ingenious Devices* (ar: كتاب الحيل (Kitab al-Hiyal) or pe: كتاب ترفندها (Ketab tarfandha)) by the brothers Banu Musa published in AD 850, which describes various automata, including primitive control methods, automatic fountains, mechanical music machines, and water dispensers. Similarly, in the court rooms around the world, various automata in the shape of singing animals were being developed and maintained. Famous examples can be found in the (now destroyed) palaces of the Khanbaliq of the Chinese Yuan dynasty and the court of Robert II, Count of Artois.

The interest in the development of automata continued into the Renaissance with the development of life-size automata, such as *The Flute Player* by the French engineer Jacques de Vaucanson (1709–1782) in 1737.

Fig. 1.1 Automation engineering in the time of the ancient Greeks: (left) Æolipile (steam engine) and (right) Automated opening device for a temple door

With the rediscovery of steam engines, the ability to develop large and complex automated system become a reality sparking the first Industrial Revolution (1760–1840). One of the first such examples was the Jacquard loom that could be programmed using punch cards to automatically weave different designs. The development of advanced systems required methods of controlling them, so that explosions and damage were minimised. The first dedicated device for controlling a steam engine, called the Watt governor, was developed by James Watt (1736–1819). The Watt governor, shown in Fig. 1.2, regulates the amount of fuel entering into the engine using the centrifugal force of the two balls. The faster the engine rotates, the further the balls are pushed outwards, which closes the throttle valve. This reduces the amount of fuel flowing into the engine, which results in a lower speed. The development of the first Watt governor was followed by a rash of patents trying to improve various aspects, including one by William Siemens (1823–1883). It was not until 1868, when James Clerk Maxwell (1831–1879) provided a mathematical description of the governor in his aptly named paper *On Governors* that a rigorous mathematical foundation for the development of methods for controlling a process was available. As the systems became more complex in the succeeding decades and centuries, there came an ever-increasing need for understanding these intricate systems in order that they be properly run so as to avoid unsafe operating conditions.

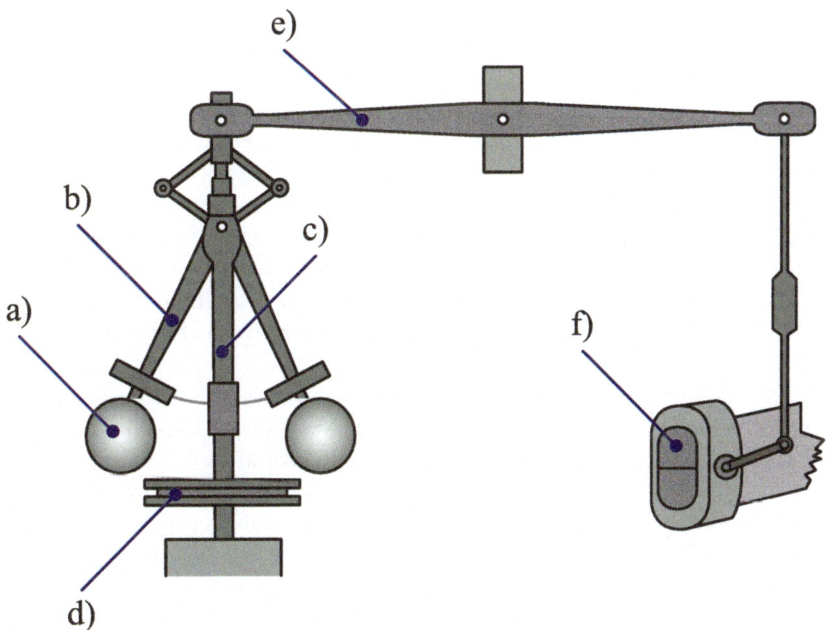

Fig. 1.2 Watt governor (a: fly ball, b: arms, c: spindle, d: sleeve, e: ball-crank lever, and f: throttle)

The demand for automation continued with the onset of the second Industrial Revolution (1870–1915), which focussed on the development of efficient manufacturing methods (production lines, Taylorism, and similar ideas) coupled with the development of electricity and the first electrical devices.

By the 1950s, a new industrial revolution, often called the third Industrial Revolution or the Digital Revolution, had started. This revolution focused on the implementation and use of complex electrical circuits that can be used to quickly and efficiently perform complex calculations. With the development of these circuits, it became easy and cost-effective to implement automation in a wide range of different fields. From the perspective of automation engineering, the key event was the development of programmable logic controllers (PLCs) that could be used to implement advanced control methods in an industrial setting. The first PLCs were developed by Bedford Associates from Bedford, Massachusetts, USA based on a white paper written by the engineer Edward R. Clark (1928–1999) in 1968 for General Motors (GM). One of the key people working on this project was Richard E. "Dick" Morley (1932–2017), who is often considered the father of the PLC. Other important work was performed by Odo Josef Struger (1931–1998) of Allen-Bradley in the period 1958–1960.

At the same time, there was an explosion of interest in a theoretical perspective on automation engineering, especially in the areas of control and process optimisation. This research by such people as Andrei Kolmagorov (ru: Андре́й Никола́евич Колмого́ров, 1903–1987), Rudolf Kálmán (hu: Kálmán Rudolf Emil, 1930–2016), and Richard E. Bellman (1920–1984) lead to a strong foundation for the subsequent development and implementation of advanced control methods in industry.

Within the context of the third Industrial Revolution, the concept of robots was also considered. The word *robot* itself was first used by the Czech author Karel Čapek (1890–1938) in his 1920 drama *R.U.R.* (cz: *Rossumovi Univerzální Roboti* or en: *Rossum's Universal Robots)* as a word for artificial humanoid servants that provided cheap labour. Karel credited his brother painter Josef Čapek (1887–1945) with inventing this word. The word *robot* stems from the Czech root *robota*, that means *serfdom*, which ultimately comes from the proto-Indo-European word, $*h_3erb^h$– that means "to change or evolve status" from which the English word *arbitrate* is also derived.

More recently, with the growth of interconnectedness and the development of smart technologies, some have proposed that a new industrial revolution is dawning. This revolution has been called the fourth Industrial Revolution, or Industry 4.0, which focuses on the development of self-functioning, interconnected systems in an increasingly globalised world. Its main drivers are ever-increasing automation and digitalisation of the industrial plant combined with globalisation and customisation of the supply chains.

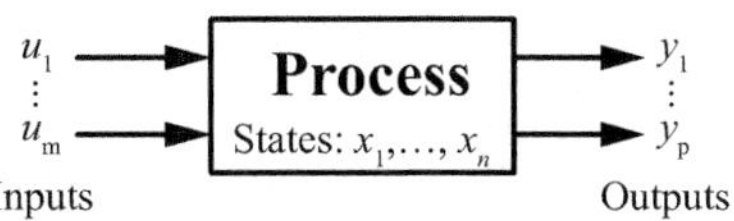

Fig. 1.3 A process in automation engineering

1.2 The Key Concepts in Automation Engineering

Automation engineering can be applied to a wide range of different cases. Instead of considering each situation separately, automation engineering has developed a set of abstract concepts that allow the ideas to be applied to any relevant situation.

The basic concept in automation engineering is a **process** or **system**.[2] Figure 1.3 shows a typical process with some important components. In automation engineering, as shown in Fig. 1.3, a process consists of **inputs**, denoted by $\boldsymbol{u}$, and **outputs**, $\boldsymbol{y}$. Inputs represent those variables, whose change will lead to a change in the process. Inputs can be divided into two types: **manipulative** and **disturbance**. A manipulative input is an input, whose value can be changed through some device, for example, the flow rate in a valve can be manipulated by opening or closing a valve. A **disturbance input** is an input, whose value cannot be changed (at least in context of the given situation), for example, the ambient temperature cannot be readily modified. Similarly, outputs can be divided into two categories: **observable** or measurable and **unobservable** or unmeasurable. Observable or measurable outputs are those variables whose value can be measured or inferred using some device, for example, the temperature of a fluid can be measured or determined using a thermometer. Unobservable or unmeasurable outputs are those variables that cannot be measured (at least in context of the given situation), for example, the density of a complex mixture consisting of multiple phases and components may not be easy to measure. Finally, we can consider the **states** of the process, which are the internal variables that describe the behaviour of the process. In many cases, the states of a process are equivalent to the outputs and can be categorised using a similar terminology. Often, not all the states can be observed.

It goes without saying that if we wish to understand the process, we require a model of the process. A model of the process is a mathematical representation of the relationship between the inputs, outputs, and states. The complexity of the model required depends on the purposes for which the model will be used. Modelling a process is a complex endeavour that requires insight into the process and the ability to handle large data sets quickly and effortlessly.

In order to complete the picture regarding the process, it is necessary to extend our view to include parts that allow us to interact with the process or influence its behaviour. Such a view is provided in Fig. 1.4, which shows the key components

[2] *Process* and *system* are practically speaking synonymous. In this textbook, *process* will be used to refer to the actual physical part, while *system* will refer to ensemble of the process and any sensors, actuators, and controllers. Obviously, a system without any sensors, actuators, or controllers will be equivalent to the process.

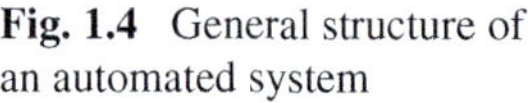

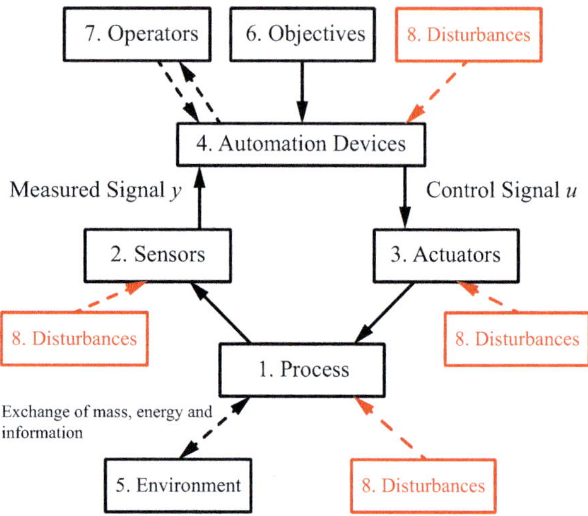

Fig. 1.4 General structure of an automated system

and their interactions. It is important to briefly consider their effect in order to understand how the process can be influenced and how it will react. In this view, the eight key components can be described as (the number is the same as in Fig. 1.4):

(1) **Process**: This represents the process under consideration. Normally, the process itself is unknown. Instead, a model of the process is used.
(2) **Sensors**: The sensors provide the ability to measure the process and understand how the variables are changing.
(3) **Actuators**: The actuators allow the value of a variable to be changed. If we cannot change the value of a variable, then it is hard to use it in an automation strategy. When deciding on which actuators to use, it is important to consider such factors as the variable being manipulated, the automation requirements (*e.g.* required accuracy, tolerance, or precision), and the type of service required (*e.g.* continuous, discrete, or emergency).
(4) **Automation Devices**: The automation devices are the controllers and related components that are used to automate the process. Most of the time the automation hardware consists of computers and other digital devices, such as **programmable logic controllers** (PLCs), that implement the required functions. The design of the automation hardware (and software) requires knowledge of the limitations and requirements of the process.
(5) **Environment**: The environment represents everything that surrounds the process to be automated that can have an effect on the overall performance of the process. This influence can be caused either by changes in other processes that interact with the process of interest or by direct environmental changes, for example, changes in the ambient temperature. The process will exchange mass, energy, and information with its environment.

(6) **Objectives**: The automation objectives play a very important role in the development and implementation of the final automation system. Poorly defined or unclear objectives can make achieving the project difficult if not impossible. Furthermore, the objectives often need to be translated from the language of business into actually implementable objectives on a system. This translation can cause additional uncertainties and lack of clarity, and thus, requires a good agreement.

(7) **Operators**: Although the human component is often minimised or ignored when designing automation systems, it is in fact very important. Many complicated automation systems have failed due to a lack of proper consideration of the operators. In general, the operators need to have the required information easily available (*no* fancy graphics are needed) and to be able to enter the required information into the system quickly and efficiently. Appropriate feedback and safety checking of the entered values must be performed to avoid problems. The operators interact with the process using a **human–machine interface (HMI)**. An HMI provides two key functions. It allows the operators to see the important process values and to manipulate as necessary the process values. Manipulating the values implies that the operators can change at what value the process operates, for example, changing the flow rate in a pipe. Finally, when designing the HMI, it is important to consider any safety features, such as logic, that limits which values can be entered by the operators. This prevents mistakes both accidental, such as mistyped numbers or the wrong information in a given field, and malicious, such as changes introduced by illegal access to the system, from having an effect on the system.

(8) **Disturbances**: Disturbances are everything that can affect the system, but whose presence cannot be directly controlled. Disturbances can originate in the environment (for example, the ambient temperature) or in the devices themselves (for example, measurement noise in sensors). One of the objectives of the automation system is to minimise, as much as possible, the effect of these disturbances on the process.

The final aspect that needs to be considered is **safety**. The automation system that has been designed should allow the process to operate in a safe manner without any unexpected behaviour. The system should also be **robust**, that is, minor changes in the process should not cause the whole system to fail catastrophically. A robust system can handle small changes in the conditions and still attain the required goals.

1.3 Framework for Automation Engineering

When solving an automation engineering problem, the following steps should be considered:

(1) **Modelling of the Process**, which involves developing an appropriate model of the process.
(2) **Analysis of the Process**, which involves using the model to analyse how the process will behave under different conditions. This can be performed using either mathematical analysis or simulations.
(3) **Design of the Automation Strategy**, which based on the properties of the process and required behaviour, provides an appropriate automation strategy that attains the required automation objectives.
(4) **Validation of the Proposed Automation Strategy**, which validates the proposed automation strategy using the process model. This determines if the proposed strategy can, in fact, achieve all the desired automation objectives. Should it be found that the proposed strategy is lacking, then the strategy needs to be refined and retested. This implies that there may need to iterate until the final automation strategy is found.
(5) **Implementation and Commissioning of the Proposed Automation Strategy**, which involves the implementation of the strategy on the real process. Naturally, the implementation on the real process may lead to changes in the strategy. This implies that before commissioning, the automation strategy should be tested on the real process in as realistic conditions as possible. In certain cases, this may not be feasible and advanced simulations, using **hardware-in-the-loop** methods can be implemented to provide a realistic simulation of the process.

1.4 The Automation-Engineering Pyramid

The **automation-engineering pyramid** is an overall description of the way in which different automation strategies can be organised and structured. It examines two key components: speed of response, or how often the system is expected to respond to changes, and process details. Figure 1.5 shows the general pyramid that consists of 6 different levels (from top to bottom):

- **Level 5—Enterprise Resource Planning (ERP) Level**: This level focuses on the abstract analysis of the overall company strategies for dealing with the current market conditions. This level runs on a very long-time horizon, often in terms of years. At this level, the focus is on market analysis, strategic investment and personnel planning, and corporate governance.
- **Level 4—Manufacturing Execution Level**: This level considers the specific strategies that allow for the overall plant/process to run. This includes such details as production planning, production-data acquisition, organisation of delivery orders, and deadline monitoring. The time horizon for actions can

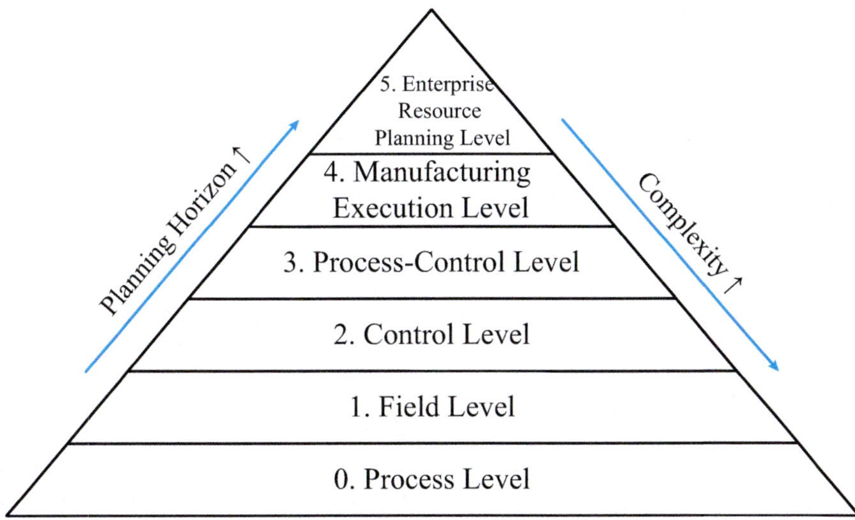

Fig. 1.5 Automation-engineering pyramid

reach into months. Often, at this level, various software systems, such as a manufacturing-execution system (MES) or a management information system (MIS), are used.

- **Level 3—Process-Control Level**: This level focuses on short-term production planning, quality control, and maintenance planning with an action horizon of at most days. Often, software, such as a supervisory, control, and data acquisition (SCADA) programme, will be used.
- **Level 2—Control Level**: This level considers how to run the algorithms for controlling the process using the available process information on the time horizon of minutes or even seconds. Various hardware, such as programmable logic controllers (PLCs) or industrial PCs (IPCs) are used to compute the required actions. It is on this level that this book focuses.
- **Level 1—Field Level**: This level focuses on the sensors and actuators. Information is actively exchanged with Level 2, often using fieldbuses. The time horizon is often in the range of seconds or faster.
- **Level 0—Process Level**: This level represents the actual process that is running in real time.

It can be noted that the process description becomes more abstract as we go from bottom to top. While Level 0 is very concrete and describes everything in great detail, Level 5 is very abstract and only focuses on describing the system inputs and outputs to give an overall picture of the system.

1.5 Chapter Problems

Problems at the end of the chapter consist of two different types: (a) Basic Concepts (True/False), which seek to test the reader's comprehension of the key concepts in the chapter; and (b) Short Exercises, which seek to test the reader's ability to compute the required parameters for a simple data set using simple or no technological aids.

1.5.1 Basic Concepts

Determine if the following statements are true or false and state why this is the case.

(1) A process consists of inputs, outputs, and states.
(2) Inputs are variables that influence the process.
(3) All outputs can always be measured.
(4) A disturbance is a variable whose value can be easily changed.
(5) The state of a process describes the internal behaviour of the process.
(6) Actuators are used to manipulate disturbances.
(7) PLCs are commonly used to automate the process.
(8) The environment has minimal effect on an automation system.
(9) HMIs should be easy to read and understand.
(10) An automation system that explodes periodically is a well-designed system.
(11) An automation system should fail the second the system deviates from the expected conditions.
(12) A system exchanges mass, energy, and information with the environment.
(13) Disturbances can affect sensors, actuators, and control devices.
(14) Validating a proposed automation strategy using appropriate models of the process is a good strategy to consider.
(15) Before commissioning of the automation strategy, it should be tested on the actual process in as real conditions as possible.
(16) The enterprise resource planning level focuses on controlling the process in fine detail and making decisions every millisecond.
(17) The process-control level often uses SCADA systems to implement its tasks.
(18) In the field level, sensors gather the information and transfer it using fieldbuses to the control level.
(19) A process variable that cannot be measured should be used to control the process.
(20) Safety is always an unimportant topic in automation engineering.

1.5.2 Short Exercises

These questions should be solved using only a simple, nonprogrammable, non-graphical calculator combined with pen and paper.

(21) How is that information can be created and destroyed, but matter and energy cannot be? Provide some examples of such cases.

(22) You have been assigned the task of designing an automation system for a traffic light system. Explain how you would apply the framework for automation engineering to this problem.

(23) You have been given the task of designing a large, multi-unit chemical plant. Explain how the automation-engineering pyramid could apply to this problem.

(24) You have been given the task of designing a self-driving car. Explain how you would implement the framework for automation engineering to this problem. Would you consider safety and robustness to be significant factors?

Chapter 2
Instrumentation and Signals

The foundational component of any automation system is the instrumentation, that is, the sensors, actuators, and the computer hardware that together produce a stream of values, often called a signal, that can be used for subsequent processing. Before we look at the sensors and actuators, it is helpful to understand the types of signals and how they can be generated.

2.1 Types of Signals

In automation engineering, signals can be classified using two domains: **time** and **value**. Each domain has two options: **continuous** and **discrete**. In general, a continuous signal can take any value within the set of (positive) real numbers, while a discrete signal can only take certain, specified values (for example, only natural numbers).

In the time domain, a signal is said to be continuous, if there exists a signal value for any time t, that is, the signal can be written as a continuous function of t. An example of a continuous-time signal would be the outdoor temperature, which contains a value for each possible time instance.

A signal is said to be discrete in the time domain, if it is only defined at certain values t_k, where $k \in \mathbb{Z}$ (or $\mathbb{N}$). Normally, it is assumed that the values are available with a specified constant sampling rate, t_s, so that $t_k = kt_s$. Obviously, as the sampling time decreases, then the signal approaches that of a continuous signal.

Similar to the time domain, the value domain of the signal can be classified into continuous or discrete. A continuous value for a signal implies that the signal can take any real number (subject to any physical constraints, for example, always positive or in the range [0, 1]). A continuous valued signal can be written as a continuous function that in general depends on time and any additional variables.

Y. A. W. Shardt, *Automation Engineering*, https://doi.org/10.1007/978-3-031-92394-4_2

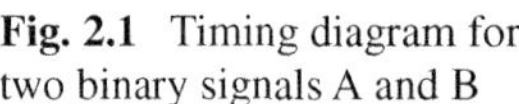

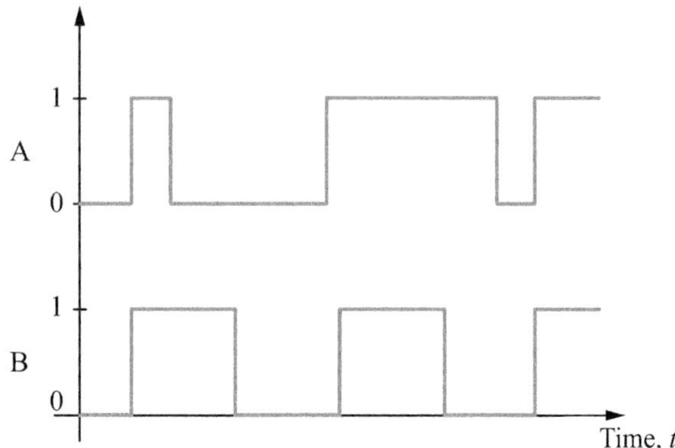

Fig. 2.1 Timing diagram for two binary signals A and B

A discrete-valued signal can only take specific values. Often the discrete values are partitioned into equidistant bands. In automation systems, a common discrete-valued signal is a **binary signal** that can only take two values (conventionally denoted by 0 or 1). Such a signal is common when dealing with alarms that are triggered when a certain condition occurs, for example, if the pressure in the reactor surpasses a given value, then the signal value is set to 1 and the alarm is triggered. Such signals work by assigning one value (say 0) to the normal state and the other value to the alarm state (say 1). Binary signals are often displayed using what is called a **timing diagram**. In a timing diagram, the binary signal is plotted as a function of time. Multiple different binary signals are normally placed on separate y-axis, but a common time x-axis. Figure 2.1 shows such a typical timing diagram.

Based on the classification in the time and value domains, it is possible to have four different types of signals, which are shown in Fig. 2.2. By convention, a signal that is continuous in both the time and value domains is called an **analogue signal**, while a signal that is discrete in both the time and value domains is called a **digital signal**.

Since most real processes require and produce continuous, analogue signals, but computers perform their computations using discrete, digital signals, there is a need to understand how signals can be converted between the two forms. Converting from analogue to digital signals is shown in Fig. 2.3, and consists of three components: **sampler**, **quantiser**, and **encoder**. The sampler measures (samples) the value of the analogue signal on a fixed frequency to convert the signal into the discrete-time domain. Next, the quantiser converts this sampled signal into the available closest value (quantum) to create a digital signal. The encoder simply encodes the digital signal into a given digital representation that can be used by the computer. This process is often denoted as an analogue-to-digital (A/D) converter.

When quantising a continuous signal, its values are compared against fixed (equidistant) quantisation levels. If the value of the continuous signal is between two decision thresholds, the lower value is usually selected. For example, if a continuous signal has a value of 0.44 with quantisation levels at 0.25 and 0.5, the quantised (sampled) signal will be set to 0.25. The selection of appropriate quantisation levels

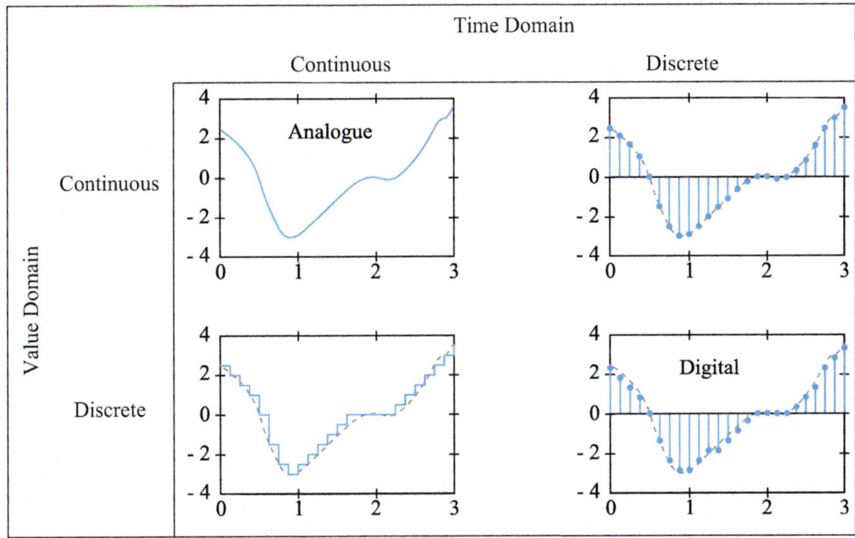

Fig. 2.2 Continuous and discrete signals

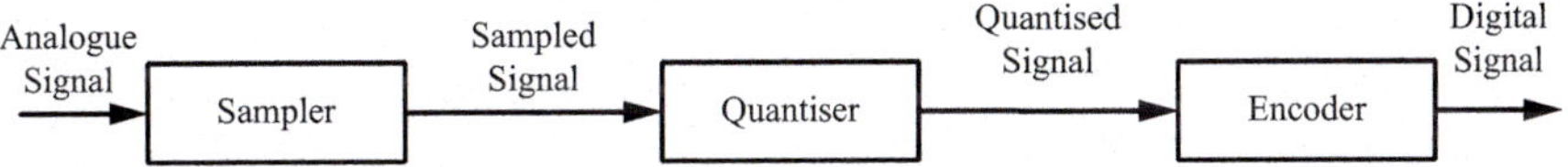

Fig. 2.3 Analogue-to-digital conversion

is important, since this selection affects the accuracy of the mapping of the process, for example, if the steps are too far apart, important information can be lost.

When going in the other direction, it is normal to assume that the value of the signal will not be changed and only the time component needs to be made analogue. This is normally accomplished using a hold, which holds the value of signal until a new value is received. The most common hold is the zero-order hold, which simply holds the last value received until a new value is received. A more accurate hold is the first-order or linear hold, which uses an interpolation between the previous two data points to obtain a linearly varying value over the sampling period. This process is called digital-to-analogue (D/A) conversion.

2.2 Sensors

A **sensor** is a device that can detect changes in a variable and upon calibration display these changes in a manner that can be understood by others, often with reference to some absolute scale.

A sensor is characterised by two properties: **accuracy** and **precision** or **reproducibility**. Accuracy measures the ability of a sensor to give the "true" value, which is usually determined based on some standard. The difference between the true and measured values is often called **bias**. Precision or reproducibility measures the variability of the sensor when measuring the same value. Ideally, it is desired that the values reported by a sensor be tightly located about the mean value, that is, the variance of the sensor values should be small. It should be noted that an inaccurate sensor may be very precise with a tight distribution about an incorrect value. Another issue to consider is the **range** of the sensor. The range of a sensor is defined as the difference between the largest and smallest values that the sensor can measure. The accuracy and precision of a sensor is often a function of the range of values that the sensor is meant to measure. The larger the range the less precise the values will be. Similarly, the smaller the range, the more precise the values will be. This implies that when dealing with processes that have values covering a large range it may be necessary to install multiple sensors which are accurate in a limited region and then using only the appropriate sensor value.

Sensors need to be calibrated before being used or to check that they are behaving as expected. Calibration involves using standards with accurate and well-defined values to compare against the measured value given by the sensor. Plotting the measured and true values against each other will allow the calibration of the sensor to be determined. A typical calibration curve is shown in Figure 2.4. There are two parameters of interest here: the y-intercept, which gives the bias, and the slope of the line (or lack of linearity). The slope of the line and the general distribution of the points suggest whether or not an appropriate calibration curve was used. Ideally, the line should be linear and the slope equal to exactly one.

A single sensor depending on its calibration and physical arrangement can be used to measure different physical variables, for example, a differential-pressure cell can measure both flow rates and level.

Selecting an appropriate sensor depends on the following criteria:

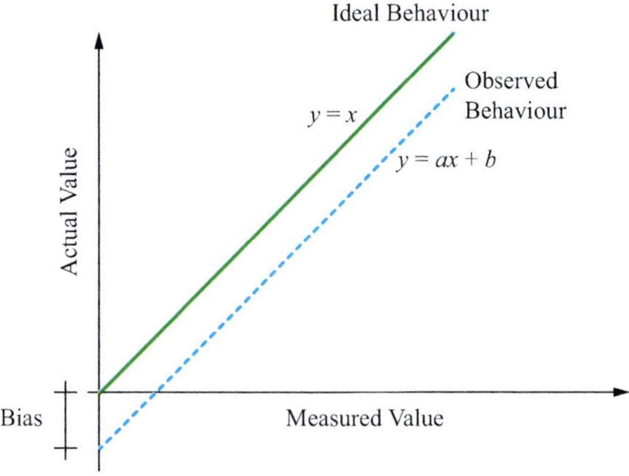

Fig. 2.4 Typical calibration curve

(1) **Measurement Range (Span)**: The required measurement range for the process variable should lie within the instrument's range.
(2) **Performance**: Depending on the specific application, different factors, such as accuracy, precision, and speed of response, will need to be considered.
(3) **Reliability**: How well does the sensor work in the given operating conditions, for example, if the sensor must be placed in harsh operating conditions, can it handle them and for how long?
(4) **Materials of Construction**: Depending on the application, the required materials of construction for the sensor may be different, for example, a temperature sensor in a blast furnace will require a different material than a temperature sensor in the living room of a house.
(5) **Invasive or Noninvasive**: An invasive sensor comes in direct contact with the object being measured, for example, inserting a probe into a liquid to measure the temperature. If an invasive sensor comes into contact with the process, it can influence the process or be influenced by the process itself. Thus, invasive methods can have issues with long-term accuracy due to fouling or corrosion of the probe surface. On the other hand, noninvasive sensors do not come in contact with the process. In such cases, the process is not disturbed, but the measurement may be less accurate. However, noninvasive sensors are generally easier to use and can be easily retrofitted into an already built environment.

2.2.1 *Pressure Sensors*

In general, most industrial pressure sensors are constructed using a **transducer** that can convert the force per given area (pressure) into an electric signal. Nonelectronic pressure sensors, often called **manometers**, do not produce an electric signal and often have a calibrated faceplate that allows the value to be read. For this reason, the signals produced cannot, in most cases, be used for industrial automation. A pressure sensor can either measure an absolute pressure or a pressure difference. A simplified schematic of these two possibilities is shown in Fig. 2.5. Most pressure sensors measure a pressure difference, which is often expressed as deviations from atmospheric pressure. As shown in Fig. 2.5a, this is accomplished by leaving one of the two taps or sides of a pressure sensor open to the atmosphere. Such a reading is often called **gauge pressure**. A negative gauge pressure is expressed by using the term **vacuum**, for example, a reading of 10 kPa vacuum, would imply that a pressure of 10 kPa below atmospheric pressure has been achieved. Instead of using the varying atmospheric pressure, it is possible to fix the pressure value on one side of the sensor and use the resulting system to measure the system pressure. As shown in Fig. 2.5b, such a set up will allow the measurement of absolute pressure.

Most transducer-based pressure sensors use some type of electric circuit to measure the strain induced by the pressure on the system. Common strain gauges include piezoresistive, capacitive, electromagnetic, piezoelectric, and

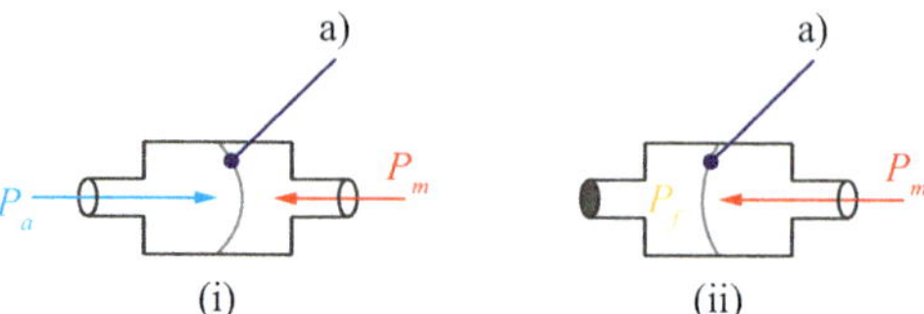

Fig. 2.5 Measurement set-up for pressure sensors: (i) Measuring differential pressure and (ii) Measuring absolute pressure (a: flexible membrane, P_a: ambient pressure, P_m: the pressure that is to be measured, and P_f: fixed pressure)

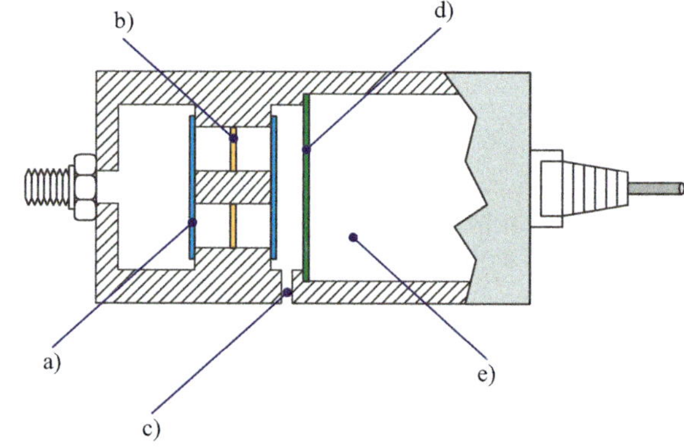

Fig. 2.6 High-pressure transducer-based pressure sensor (a: measuring diaphragm, b: strain gauge, c: reference hole to the atmosphere, d: true gauge diaphragm, and e: area for the resistor for temperature correction and internal electronic amplifiers)

optical. Figure 2.6 shows a typical high-pressure transducer-based pressure sensor. Similarly, manometers are also based on the effect of pressure on some system property. They include the common hydrostatic manometers, which basically measure the difference in pressure between the two taps, and mechanical manometers, which measure the effect of pressure on the strain of the material. Mechanical manometers have the advantage that they do not interact strongly with the fluid and can provide very sensitive readings. On the other hand, compared to the hydrostatic manometers, they can be more expensive.

2.2.2 *Liquid-Level Sensors*

Determining the liquid level in tanks or other similar containers is a very common need in many chemical plants. Liquid level can be determined using many different methods, including differential-pressure cells, floats, and various radio-based methods. Of these, the most common approach is to use a differential-pressure cell, which measures the pressure gradient between the top of the liquid and the bottom of the liquid. Since pressure is proportional to the height of the liquid in the tank, it is relatively easy to calibrate and determine the height. On the other hand, since density depends on the temperature, this approach will not work in cases where there can

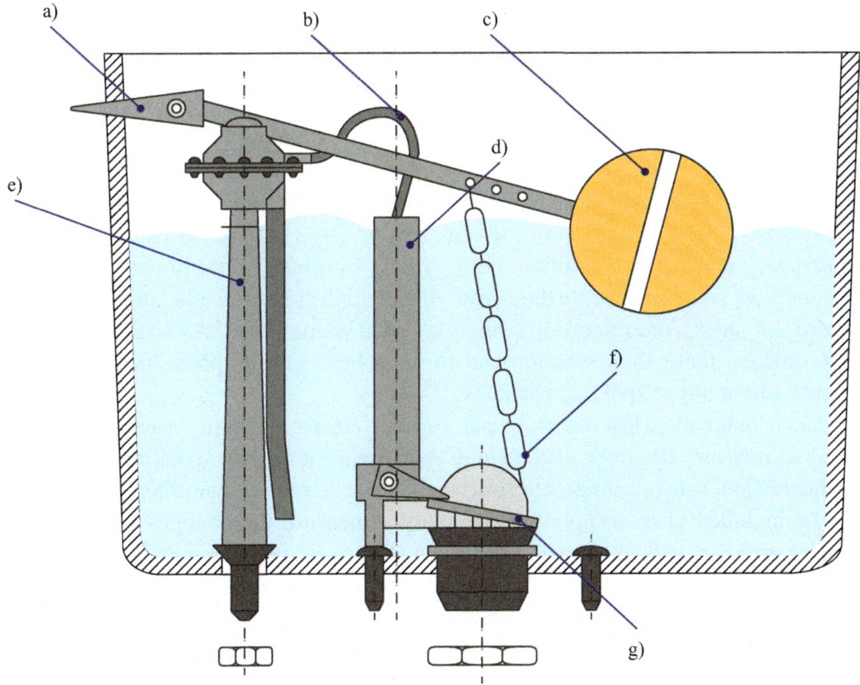

Fig. 2.7 Level measurement and control using a float (a: trip lever, b: refill tube, c: float, d: overflow tube, e: ballcock, f: lift chain, and g: flush valve)

be wide fluctuations in temperature (for example, a boiling liquid). A float measurement device works on a similar principle, of pressure difference, but records the values differently. Figure 2.7 shows an example of float to measure the level in a toilet for controlling the flow of water into the toilet tank. This shows a relatively simple example of automation that can be implemented with an appropriately selected sensor. Finally, various radio-based methods, for example, ultrasonic pulse generating devices, can be used to determine the surface level. However, in order to get an accurate estimate, it is required that the surface be relatively flat and consistent. Froths and other particles can affect the accuracy and precision of the measurements.

2.2.3 *Flow Sensors*

Flow sensors are used to measure the speed at which a liquid or gas is moving. There are three main types of flow sensors: **mechanical flow sensors**, **pressure-based flow sensors**, and **electromagnetic flow sensors** (including optical and ultrasonic flow sensors). Depending on the fluid present, each of the three types will have different accuracies and characteristics.

Mechanical flow sensors are based on the idea of timing how long it takes the liquid or gas to fill some known unit of volume. Most mechanical flow sensors have some type of wheel or paddle that is turned by the flowing medium inducing an electric signal, which is then calibrated to give a flow rate. In general, mechanical flow sensors are good with simple fluids, for example, water in a single phase over a limited range of flow rates. Suspended particles in the fluid as well as multiple phases can cause the mechanical flow sensor to give incorrect readings.

Pressure-based flow sensors measure the pressure difference caused by some constriction in flow to determine the flow rate. Common pressure-based flow sensors include Venturi tubes, orifice-plate differential-pressure cells, and Pitot tubes. Figure 2.8 shows the operating principles of a Venturi tube. As with mechanical flow sensors, these flow sensors tend to work best with uniphase flow of simple fluids without any suspended particles.

Electromagnetic flow sensors use various electromagnetic waves (including light) to measure the flow of the fluid. Although not strictly speaking an electromagnetic flow sensor, ultrasonic flow sensors are based on a similar principle and can be included here. Magnetic flow sensors measure the changes induced by a flowing fluid in a local electric field. In many cases, the flowing fluid should be conductive and the surrounding pipe a nonconductor. Optical flow sensors use the time it takes for small, suspended particles inside a flowing gas to cross two laser beams. Based on this time, the bulk flow rate can be computed. Finally, ultrasonic flow sensors use the Doppler effect to measure the flow rate. An advantage of the ultrasonic flow sensors is that they can be nonintrusive, that is, the measurement

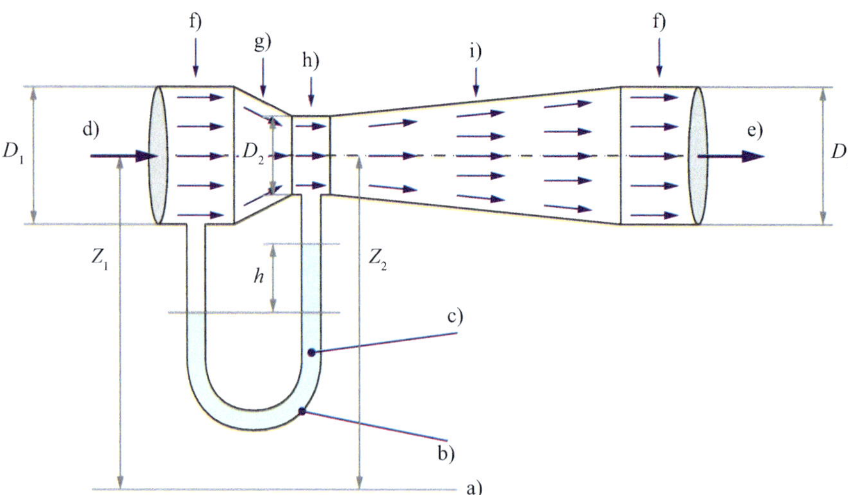

Fig. 2.8 Venturi tube (a: datum, b: U-tube manometer, c: manometer fluid, d: inlet, e: outlet, f: main pipe, g: converging cone; h: throat, i: diverging cone, D_1: diameter of the main pipe, D_2: diameter of the throat, Z_1: reference height 1, Z_2: reference height 2, and h: height difference in the manometer)

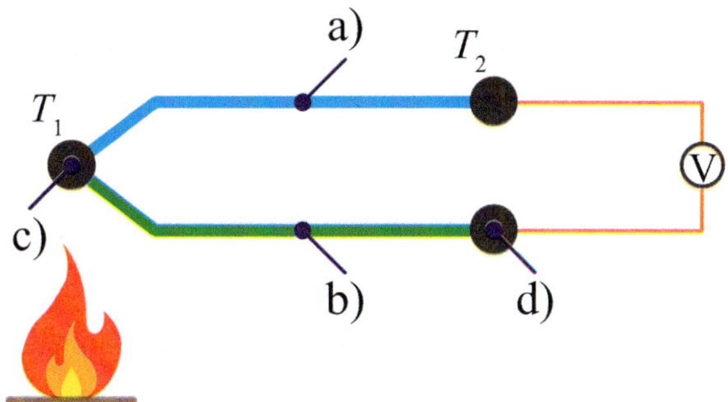

Fig. 2.9 Thermocouple (a: metal 1, b: metal 2, c: measurement point, d: reference location, T_1: temperature to be measured, T_2: reference temperature, and V: voltmeter)

can be done without coming in contact with either the pipe or the fluid. On the other hand, these sensors require the speed of sound for the flowing fluid to be known in order to calibrate the results correctly.

In most industrial cases, pressure-based flow meters are used, since they are simple to use, robust, and easy to maintain. On the other hand, when dealing with exotic or extreme fluids, then more complex methods may be required.

2.2.4 Temperature Sensors

Temperature sensors are used to measure the temperature of a stream, be it a liquid or gas. The most common industrial temperature sensor is a **thermocouple**, which uses the Seebeck effect to induce a voltage that depends on the temperature difference. The Seebeck effect describes a phenomenon where a temperature difference between two different electrical conductors or semiconductors creates a voltage difference between the two materials. It is possible to calibrate this voltage change with temperature and hence use it. The construction of the thermocouple is shown in Fig. 2.9. Determining an appropriate span and calibration for a thermocouple is very important in order to allow for maximal benefit. Inappropriate calibration can lead to accuracy issues. The most common types of thermocouples are shown in Table 2.1.

2.2.5 Concentration, Density, Moisture, and Other Physical-Property Sensors

There exist various online sensors that can quickly provide readings for many different types of physical properties. Unfortunately, most of these sensors are relatively limited in their accuracy or precision. This often results from the extreme or nonideal conditions in which such systems are used. Often the laboratory measurements are more accurate and trusted. Most of these sensors use various photonic, magnetic, or sonic-pulse methods to determine concentrations or densities.

Table 2.1 Types of thermocouples

Thermocouple type	Temperature range (°C)	Material of construction	Comments
K	−200 to 1350	Chromel−alumel	Cheap, not too precise
E	−110 to 140 (narrow) −50 to 740 (wide)	Chromel−constantan	Good for cryogenic use
J	−40 to 750	Iron-constantan	More sensitive than K
B and R	50 to 1800	Platinum-rhodium alloys	Good for high temperature use, expensive
C, D, and G	0 to 2320	Tungsten/rhenium alloys	Cannot be used in presence of oxygen, expensive.
Chromel-gold/iron	−273.15 to 25	Chromel-gold–iron	Cryogenic applications

Consider, for example, a moisture sensor for water in soil. At present, there are 4 ways to measure the moisture: tensiometers, which measure the water tension in the soil; electrical resistance blocks, which measure the resistance of a ceramic block in contact with the soil; electrical conductivity probes, which measure the conductivity of the soil between 2 plates; and dielectric sensors, which measure the dielectric constant of the soil. All methods require regular maintenance and are sensitive to ambient conditions, for example, placing more fertiliser will change the electrical conductivity of the moisture and hence disrupt the sensor.

2.3 Actuators

An **actuator** is a device that can change the amount of some variable that enters a system. An actuator is characterised by three properties: **accuracy**, **precision** or reproducibility, and **performance**. Accuracy measures the ability of an actuator to give the "true" value, which is usually determined based on some standard. The difference between the true and measured values is often called **bias**. Precision or reproducibility measures the variability of the actuator when delivering the same value. Ideally, it is desired that the value delivered by an actuator be close to the desired true value. The final issue to consider is the performance or how long does it take for the actuator to respond to a change. In general, it is desired that an actuator respond quickly to changes and deliver the final desired value in a short period of time. Most actuators are designed so that 0% corresponds to no value and 100% corresponds to the maximal value. A common problem with actuators is that their response is nonlinear. One way to resolve this problem is to use a **characterisation** function that will convert the desired linear values into the actuator's nonlinear values.

Actuators need to be calibrated before being used or to check that they are behaving as expected. Calibration involves using standards with accurate and well-defined values. However, the exact relationships and behaviour of the calibration will depend on the specific actuator.

There are three common actuators: **valves**, **pumps**, and **variable current actuators**.

2.3.1 Valves

Valves are one of the most common actuators seen in a plant. They allow the flow rate of a liquid or gas to be controlled. Since valves are so ubiquitous in plants, there is a vast amount of work done on understanding valves and how they affect the performance of automation. The most common type of control valve is a pneumatic control valve that uses air to change the flow rate. A typical control valve is shown in Fig. 2.10. A pneumatic control valve consists of three components: the current-to-pressure converter (I/P converter), the valve itself, and a positioner. The I/P converter takes the electric 4–20 mA signal and converts it to an appropriate pressure signal that can provide the motive force required to move the piston. There exist two types of control valves: **air-to-close** and **air-to-open**. Air-to-open valves are also called **fail-close valves** because in the event of a loss of air pressure, the valve will close. Similarly, air-to-close valves are called **fail-open valves** because in the event of a loss of air pressure, the valve will remain open. The choice of air-to-open or air-to-close valves is based on the outcome of a process hazard review.

For an air-to-close valve as shown in Fig. 2.10, the compressed air enters the top of the valve actuator, exerts force against the diaphragm, and moves the diaphragm until the resistance force from the spring is equal to the force on the diaphragm from the compressed air. An increase in air pressure will tend to push the diaphragm down, while a decrease in air pressure will result in the spring forcing the diaphragm up to a new equilibrium point. The valve stem is attached to the diaphragm, and as the stem moves up and down it changes the position of a tapered plug (or "trim") relative to a seat. As the stem moves, it changes the cross-sectional area available for flow, and thus the resistance to flow. For an air-to-close valve, an increase in air pressure will push the plug towards the seat, thus reducing the area available for flow, increasing the resistance to flow, and reducing the flow rate though the valve. For an air-to-open valve, the air enters at the bottom of the diaphragm, so an increase in air pressure will raise the diaphragm, which will open the valve.

Many modern valves have an additional element called a **positioner** that seeks to overcome any potential errors in the valve. A positioner basically compares the current valve location against the reference value and will change the air supply to allow for the difference to be zero.

The behaviour of a valve is normally specified based on the percentage of the total distance that the valve is opened or closed. This eliminates the need to know

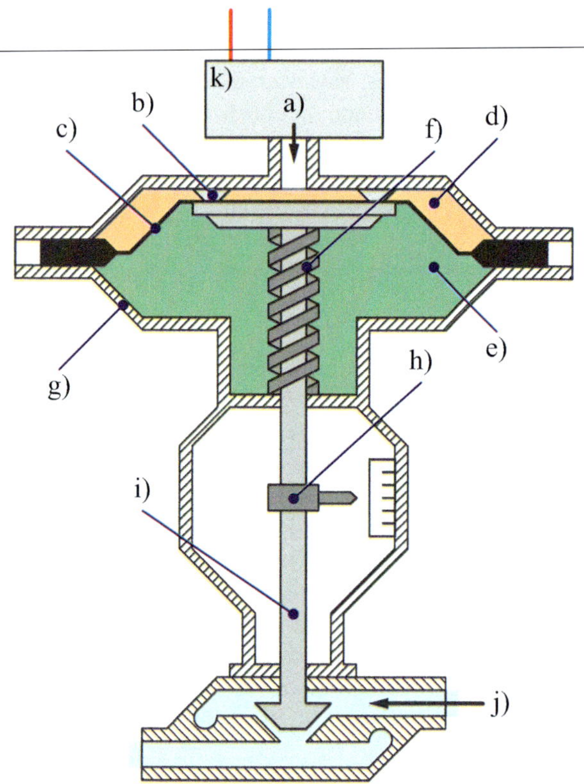

Fig. 2.10 Pneumatically actuated control valve (air-to-close) (a: supply air connection, b: mechanical stop, c: diaphragm, d: upper chamber, d: lower chamber, f: spring, g: housing, h: local position indicator, i: stem, and j: flow direction, and k: transducer)

the exact flow rates. Therefore, the flow rates of a valve are often stated in terms of percent open (often abbreviated as %open).

Properly selecting a valve for automation (or control) is very important. There are two variables to consider: sizing and dynamic performance.

Valve Sizing

Control valves must be specified, just like piping, heat exchangers, and other process equipment. The sizing of a control valve determines the range of flow rates over which the valve can produce a hydrodynamically stable flow.

A valve that is too small will not permit enough flow when it is fully open, which is defined in terms of the needs for process regulation. A control valve must permit significantly greater flow than the steady-state requirement in order to be able to provide reasonable performance. A valve that is too large will have a sufficiently high maximum flow but will provide poor regulation when the flow is low. For a number of reasons, valves are not very precise instruments, and the relative errors tend to be greatest when the valve is almost closed.

A valve that is open by less than 10%open is generally considered effectively closed, and a valve that is more than 90%open is generally considered effectively fully open.

To determine whether a control valve is undersized or oversized, examine the range of controller output values used when the valve is in service. If the valve spends a significant fraction of the time fully open, then it can be considered undersized. If it spends a significant fraction of the time less than 10%open, or if the flow through the valve reaches a maximum before the valve is fully open, then the valve is oversized.

Dynamic Performance of Valves

In most flow control applications, it is desirable for the flow through the valve to be a linear function of valve position. If a plot of valve %open as a function of steady-state flow rate produces a straight line, then the valve is said to be linear. There exist two types of valves which will give a nonlinear plot: **quick-opening valves** and **equal-percentage valves**, whose plots are shown in Fig. 2.11. A quick-opening valve, as its name suggests, will quickly open and reach the maximal flow rate, while an equal-percentage valve will reach the maximal flow rate more slowly. Since this behaviour of the valve is determined by the manufacturer, it is often called the **inherent valve characteristic**. If the valve is nonlinear, then the control performance can be improved by including a characterisation block that converts the linear flow rates that are desired into the corresponding percentage, rather than assuming a linear relationship.

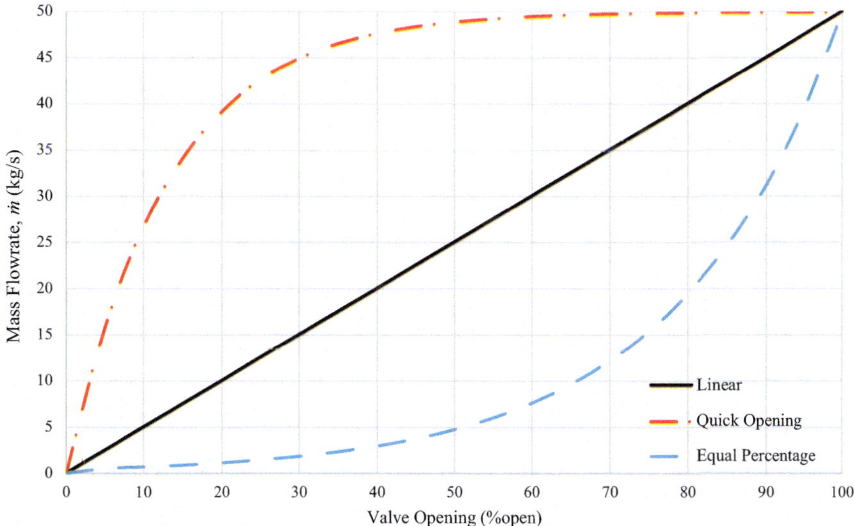

Fig. 2.11 Inherent valve characteristics

Furthermore, since a valve is a mechanical system, friction will cause two additional types of nonlinearities to be present in the system: **static** and **dynamic**. Static nonlinearity refers to the nonlinearity introduced by static friction when initiating changes in the valve position, while dynamic nonlinearity results from the dynamic friction in the valve, usually between the valve stem and the seal. Static friction between the stem and seal results in the valve not moving when there is a small change in the control signal (air pressure) input to the valve. Static friction, or "stiction," can significantly affect controller performance.

To find static and dynamic nonlinearities in a control valve, step the valve from 0%open to 100%open and then back to 0%open in small steps and record the steady-state flow at each step. Plot the data points on a graph of steady-state flow versus %open, and use different lines for opening and closing. The resulting graph will be similar to that shown in Fig. 2.12, where the ideal valve behaviour is shown as a dashed line. Static nonlinearity consists of the **deadband** region over which the flow rate does not change even though the signal to the valve increases and the **slip-jump** behaviour, while dynamic nonlinearity consists of **hysteresis**, which is the gap between the two lines. Slip-jump behaviour is caused by static friction that must first be overcome before the valve changes its position. Since dynamic friction is much smaller than static friction, the valve will overshoot the position defined by the force as soon as it starts to move. This will make the curve look like a staircase. Hysteresis arises from the difference between the measured flow rate when opening and closing a valve. This effect can be explained by the different effect of dynamic friction on the moving valve. When the valve is being opened, due to dynamic friction, it will open less than desired. On the other hand, when it is being closed, the valve will be more open than specified, that is, it will close less than desired.

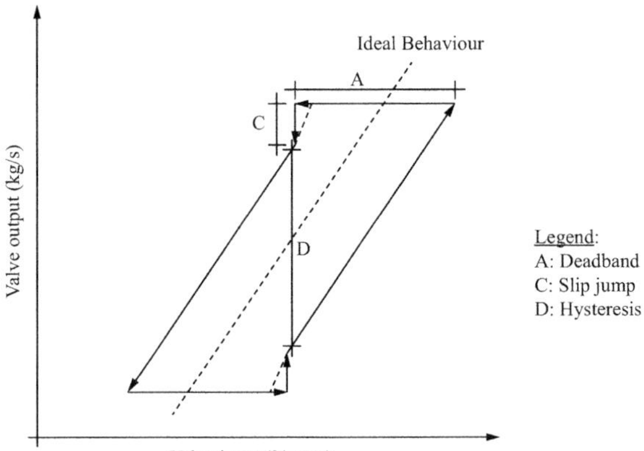

Fig. 2.12 Phase plot for the typical behaviour of a valve with stiction (after Shoukat Choudhury et al. (2005)). The arrows show the direction in which the values were changed

2.3.2 *Pumps*

Another actuator that can control the flow rate in a stream is a pump. A pump is a mechanical device that takes a fluid, most often a liquid, from a storage tank and moves it to some other location. The main types of pumps are centrifugal pumps, positive-displacement pumps, and axial-flow pumps. In centrifugal pumps, the flow direction changes by 90° as it moves over the impeller, while in an axial-flow pump the flow direction is not changed. In positive-displacement pumps, the fluid is trapped in a fixed volume and forced (or displaced) into the discharge pipe. The traditional hand pump is a good example of a positive-displacement pump. Of these three types, the most common type is a centrifugal pump. A positive-displacement pump is often used if flows are small or extreme precision is required. Figure 2.13 shows a schematic of a centrifugal pump, while Figure 2.14 shows a positive-displacement pump.

Compared to valves, pumps tend to provide better control of the flow rate and have fewer nonlinear characteristics. However, they have much higher energy consumption than valves. The main considerations for a pump, as for valves, are **sizing** and **performance**.

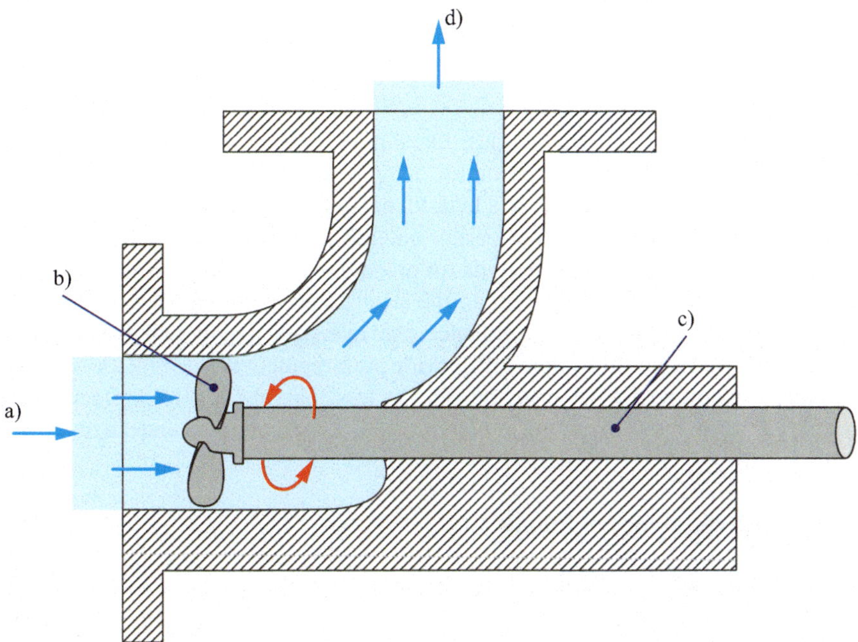

Fig. 2.13 Centrifugal pump (a: inflow, b: impeller, c: shaft, and d: outflow)

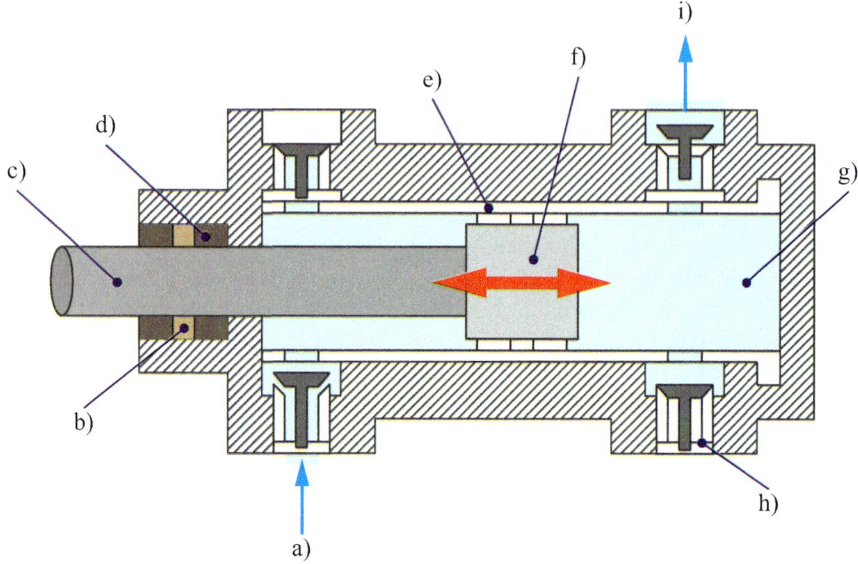

Fig. 2.14 Positive-displacement pump (a: inflow, b: packing, c: piston rod, d: stuffing-box bushing, e: liner, f: piston, g: working fluid, h: valve, and i: outflow)

Pump Sizing

Control pumps must be specified, just like piping, heat exchangers, and other process equipment. The sizing of a control pump determines the range of flow rates over which the pump can produce a hydrodynamically stable flow.

An undersized pump will not permit enough flow when it is fully operating, which is defined in terms of the needs for process regulation. A control pump must permit significantly greater flow than the steady-state requirement in order to be able to provide reasonable performance. An oversized pump will have a sufficiently high maximum flow, but will provide poor regulation when the flow is low.

Pumps operating for most of the time at less than about 10% are said to be oversized, while those operating at more than 90% are said to be undersized.

Dynamic Performance of Pumps

Unlike for a valve, the performance of a control pump is easier to quantify. In general, pumps are linear. The only significant issue is that with certain types of pumps a **deadband** may be present at low flow rates, that is, there may be no observed flow. This can be attributed to the fact that the pump cannot overcome the effects of gravity and friction, and hence, produce a flow.

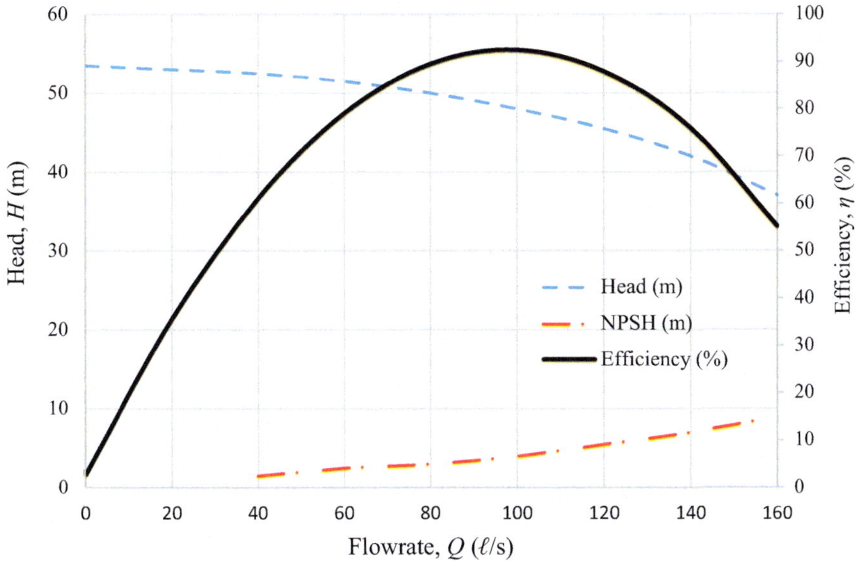

Fig. 2.15 Pump characteristic curve for a centrifugal pump

The behaviour of a pump is characterised by its **pump characteristic curve**, which is often supplied by the manufacturer of the pump. A typical pump characteristic curve for a centrifugal pump is shown in Fig. 2.15. The **head**, H, normally expressed in units of length, such as metres or feet, shows how high a given column of liquid could be lifted by a pump. It represents the effective pressure gradient that the pump can overcome. The **efficiency**, η, represents how much work put into the pump is converted into lifting the liquid. As in many engineering applications, the higher the efficiency, the better it is. For centrifugal pumps, the **net positive suction head** (NPSH) is the minimum head (pressure) at the inlet before cavitation occurs. Cavitation is defined as the boiling of a liquid in a pump, which is evidently a very undesirable event. Therefore, the head at the inlet must be greater than the specified value.

2.3.3 Variable Current Devices

The final actuator of interest is a variable current device that can modulate (vary) the current entering a device. Since these devices are mechanical, they tend not to have any issues with nonlinearities or undesired behaviour. As with all equipment, sizing can be a problem that needs to be appropriately resolved.

2.4 Programmable Logic Computer (PLCs)

Programmable logic computers (PLCs) are small computers that have been made robust and rugged. They are often used in industry for controlling processes, such as assembly lines, robots, or complex chemical processes. PLCs allow the collection of the different signals and their subsequent processing to reach a decision or action that may need to be taken. Given the computational power, they can also perform relatively advanced logical and mathematical functions that can be used to control the process.

As shown in Fig. 2.16, a typical PLC consists of 6 key components:

(1) **Inputs**: These allow information from outside the PLC to be incorporated and used. Most often, they are electrical signals coming from sensors or switches.
(2) **Power Supply**: This provides the power required to run the PLC and operate all the circuitry. Internally, most PLCs use a 5-V standard. However, the power supply most often provides 230 V AC, 120 V AC, or 24 V DC. Furthermore, the power supplies are often built as a replaceable module, so that depending on the application, the appropriate power can be provided, for example, in Europe one could use 230 V, but in North America 120 V.
(3) **Central Processing Unit (CPU)**: The CPU is the brain of the PLC that performs all the required instructions, calculations, operations, and control functions.
(4) **Memory**: The PLC must also contain memory or the ability to store relevant information for future use. The amount of memory available depends on the PLC and the programming requirements. Some PLCs can have additional memory cards inserted, so that they have more available memory. There are two main types of memory:

    a. **Read-Only Memory (ROM)**: ROM is the permanent storage for the operating system and system data. Since true ROM cannot be changed, in practice, an **erasable programmable ROM** (EPROM) is used, so that it is possible to update the operating system for the PLC.
    b. **Random-Access Memory (RAM)**: RAM is used to store all other information required by the PLC including any programmes and variables. Accessing RAM is very fast. However, when power is lost, the information in RAM is also lost. Therefore, a PLC will always have an additional battery to maintain power to the RAM, so that the information that is contained in the RAM is not lost during a power outage.

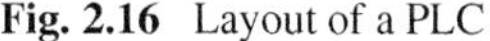
Fig. 2.16 Layout of a PLC

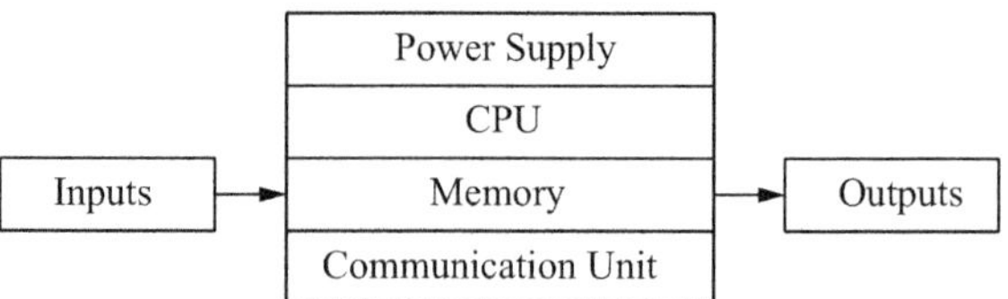

(5) **Communications Unit**: The communications unit allows for the PLC to interact with other devices to exchange information using different types of protocols. Often, this exchange of information involves sending new or updated programmes to the PLC. In general, the communications unit will often include the ability to communicate with an operator panel, printers, networks, or other computers.
(6) **Outputs**: These allow the PLC to exchange information with other devices to cause them to take an action. Such devices include motors, valves, pumps, and alarms.

In a PLC, communication between the components occurs using groups of copper wires called **buses**. A bus consists of a bundle of wires that allow for the transmission of binary information, for example, if the bus contains eight wires, then it is possible to transmit up to 8 bits of information per bus. A typical PLC consists of four buses:

(1) **Data Bus**: The data bus is used to transfer information between the CPU, memory, and I/O.
(2) **Address Bus**: The address bus is used to transfer the memory addresses from which the data will be fetched or to which data will be written.
(3) **Control Bus**: The control bus is used to synchronise and control the traffic circuits.
(4) **System Bus**: The system bus is used for input-output communications.

Since a PLC operates more or less similarly to that of a computer, this means that in order for a PLC to do anything useful, it must be programmed. The basic idea for a PLC is to monitor and control a **process**. A process in this context is anything that requires monitoring and control and can range from a simple unit that requires its output to be maintained at a prespecified value to a complex, highly interacting system like a room with multiple heating ovens, lights, ventilations systems, and windows, wherein the temperature, carbon-monoxide levels, and humidity must be monitored and maintained at safe levels. In PLC terms, a process is said to be in a given **operating mode**,[1] when the process is operating in some specified mode, for example, when the pump is on, off, or operating. Finally, the PLC requires a **user programme** that takes the inputs, outputs, and internal information to make decisions about the process.

Before we look at how a PLC operates, it can be useful to note the different operating modes (states) that a PLC itself can be in. In general, the operating modes for a PLC are **programming**, **stop**, **error**, **diagnostic**, and **run**. In programming mode, a PLC is being programmed, usually using an external device. In stop mode, the PLC is stopped and will only perform some basic operations. In error mode, the PLC has encountered some sort of problem and has stopped working. In diagnostic mode, the PLC runs without necessarily activating any inputs or outputs, allowing for the validity of the programme to be determined. Often, test signals are used in place of the real inputs and outputs. In run mode, a PLC is actively working

[1] Often also called a *state*, but this term is avoided since it has another meaning in control and process analysis.

and performing the requested actions. Each PLC manufacturer may call these operating modes different names and not all of them may be present for a given PLC.

Naturally, the most important mode from the perspective of automation is the run mode. Thus, its behaviour will be examined in greater detail. In run mode, the PLC performs the same four operations in a repeating cycle:

(1) **Internal Processing**, where the PLC checks its own state and determines if it is ready for operation. Should the response from hardware or communication units be lacking, the PLC can give notice of these events by setting a **flag**, which is an internal Boolean address (or visual indicator) that can be checked by the user to determine the presence of an error state. Normally, the PLC will continue operation, unless the error is serious, in which case, it will suspend operation. In this step, software-related events are also performed. These include such things as updating clocks, changing PLC modes, and setting **watchdog** times to zero. A watchdog is a timer that is used to prevent a programme from taking too long to execute, for example, it could be stated that if a programme does not terminate in one second, then it could be stuck inside an unending *while*-loop and an error mode will be returned.
(2) **Reading Inputs**: Next, all the input values are copied into memory. This means that the PLC will only use the values from memory rather than checking to make sure that the most recent value is available. This also means that at any given point the programme will use the same values during the same scan time. Furthermore, reading values from memory is faster than reading them each and every time from the input.
(3) **Executing Programmes**: After the inputs have been read, the programmes are executed in the order in which the code has been written. The order of execution can be changed by changing the associated **priority** of the given code or using conditional statements and subroutines. It should be noted that at this point only internal variables and output addresses are updated in memory; the physical outputs are not changed at this point.
(4) **Updating Outputs**: Once the programmes have been finished, the output memory is written so that the state and values of the outputs can be updated. This will then complete a single cycle and the PLC will return to the first step, that is, internal processing.

Since it can be seen that the PLC in run mode performs these four operations repeatedly, the question becomes what is the best approach for how the PLC should repeat these operations. By convention, a single pass through the four operations is called a **scan**. The **scan time** or **cycle time** is the amount of time required to perform a single scan. In practice, the scan time can vary between scans due to the presence of different events or conditions requiring the execution of more or less code. In order to accomplish the repetitions effectively, a programme can be associated with a **task**, whose execution type can be specified. There are three common execution types:

(1) **Cyclic Execution**: In cyclic execution, the time between scans is fixed. Obviously, the time must be set so that the PLC has the time to complete all

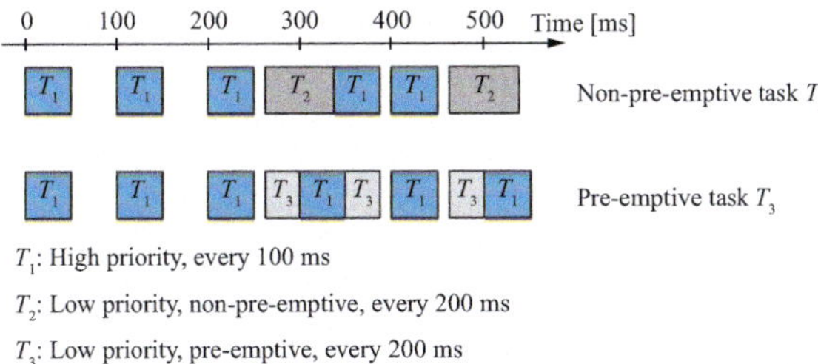

Fig. 2.17 Pre-emptive and non-pre-emptive tasks

the required code. Naturally, certain tasks, such as counting or timing, should always be run on cyclic execution.

(2) **Freewheeling Execution**: In freewheeling execution, as soon as one scan has been completed, then the next scan is started. This is the fastest way of running a task, since there is no waiting between scans.
(3) **Event-Driven Execution**: In event-driven execution, a task is only executed when a given Boolean condition is fulfilled. Event-driven execution is useful for emergency stop routines, start-up routines, and other extra-ordinary events.

Furthermore, the nature of the task must be specified: can the task be interrupted by another task? If the task can be interrupted, then it is called **pre-emptive**; if it cannot be, **non-pre-emptive**. The difference between these two types of tasks is shown in Figure 2.17. Finally, the priority of a task must be specified ranging from high to low (the exact details depend on the specific PLC and standard used).

2.5 Communication Devices

The last type of instrumentation is the **communication devices** which allow all the actuators, sensors, and control logic to communicate with each other. The following equipment is often found:

(1) **Analogue-to-Digital Converters (A/D Converter)**, which convert the analogue signal received from the sensor into a digital signal. At this point, quantisation (or resolution) can be an issue. If an insufficient number of levels (decimals) are present, then the resulting data may not carry as much information as before.
(2) **Digital-to-Analogue Converters (D/A Converter)**, which convert a digital signal received from the computer/software into an analogue signal that can be used by the actuator. Different methods exist for implementing this conversion including a zero-order hold, where the previous value is maintained until a change occurs, or a first-order hold, where some average of the previous values is used until a new value is available. In most implementations, a zero-order hold is used due to its simplicity and sufficiency.
(3) **Control Software**, which can reside either on the programmable logic computer or as a separate software programme on a computer.

(4) **Data Historian**, which stores all the values for future retrieval. Selecting an appropriate sampling time, or how fast the data are recorded, can determine the usefulness of the stored data for future applications.
(5) **Network Cables, Switches, and Accessories**, which physically connect all the equipment and allow for the strategy to be implemented.

The main issue with the design of the communication units is the available **bandwidth**. The faster the data is sampled and the more computations that need to be performed, the larger the bandwidth and computational power that will be needed.

In automation, signals can be encoded using many different standards. The two most common standards are the current-based standard of 4 to 20 mA and the pressure-based standard of 3 to 15 psig.[2] It should be noted that both of these standards do not start at zero, since a value of zero is ambiguous: is the device not working properly or is the value actually zero. Using a **live zero**, that is, the value of zero corresponds to some nonzero current or pressure makes it is possible to distinguish between the case of a zero value and a faulty device. Furthermore, a live zero allows part of the remaining current or pressure to be used for operating the device without needing an additional power supply. This means that the device can be used in remote areas without its own power supply. The lower limit for the live zero was historically determined as the smallest measurable value. In most cases, the upper limit was chosen so that the lower and upper limits are in the ratio 1:5.

2.6 Further Reading

The following are additional resources on the topics mentioned in this chapter:

(1) **Actuators and Sensors**

a. G. D. Ulrich and P. T. Vasudevan (2004). *Chemical Engineering Process Design and Economics: A Practical Guide* (Second Edition), Durham, New Hampshire, USA: Process Publishing.
b. Don W. Green (editor) (2018). *Perry's Chemical Engineers' Handbook.* (85th Edition), New York, New York, USA: McGraw-Hill.

(2) **PLCs**: K.-H. John and K. Tiegelkamp (2009). *IEC 61131-3: Programming Industrial Automation Systems: Concepts and Programming Languages, Requirements for Programming Systems, Decision-Making Aids* (Second Edition), Berlin, Germany: Springer-Verlag.

[2]A psi is a unit of pressure in the imperial system of measurement. It is an abbreviation for pounds (force) per square inch. The conversion factor is 1 psi = 6.894 757 kPa. The *g* represents *gauge* or the value above atmospheric pressure.

2.7 Chapter Problems

Problems at the end of the chapter consist of two different types: (a) Basic Concepts (True/False), which seek to test the reader's comprehension of the key concepts in the chapter; and (b) Short Exercises, which seek to test the reader's ability to compute the required parameters for a simple data set using simple or no technological aids.

2.7.1 Basic Concepts

Determine if the following statements are true or false and state why this is the case.

(1) An analogue signal is continuous in both the time and value domains.
(2) A binary signal is an example of a digital signal.
(3) A signal can only be discretised in the value domain.
(4) A precise sensor will always give a value close to the true value.
(5) If the observed sensor value is 3 kg with a standard deviation of 0.5 kg and the true value is 10 kg, then we can say that the sensor is precise and accurate.
(6) A manometer measures the pressure difference using a transducer.
(7) A differential-pressure cell can measure the level of a liquid in a tank.
(8) Venturi tubes are an example of a pressure-based flow sensor.
(9) The Doppler effect can be used to measure flow rates.
(10) The Seebeck effect allows us to measure pressure changes.
(11) A J-type thermocouple can be used to measure the temperature of molten silica which is at least 1000°C.
(12) It is not possible to use thermocouples to measure temperatures below 0°C.
(13) Actuators should be selected so that they are used in the range 20–60%.
(14) Valves are a type of actuator that restricts flows.
(15) An air-to-close valve will remain open should the air supply fail.
(16) A quick-opening valve is useful for rapid dosing of a liquid.
(17) Slip-jump in valves results from dynamic friction of the moving parts.
(18) The efficiency of a pump represents the pressure gradient that it can produce at a given flow rate.
(19) A PLC consists of inputs, power supply, CPU, memory, outputs, and communication devices.
(20) A bus in a PLC is a location in memory used to store information about where different variables are stored.
(21) A PLC in programming mode is being programmed from an external device.
(22) A flag in a PLC is a Boolean variable that shows the presence of an error state.

(23) A watchdog in a PLC is a variable that prevents the PLC from being interrupted by external users as it is running.
(24) A PLC scan is the amount of time it takes for the PLC to read the inputs.
(25) Freewheeling execution occurs when the PLC executes a task solely on demand from an external event.
(26) A pre-emptive task can never be interrupted.
(27) An analogue-to-digital converter is found in all computer-based automation solutions.
(28) A data historian stores the data collected from a process.
(29) Live zero implies that when the sensor value is zero then the current has a value of 4 mA.
(30) A PLC normally has a battery to provide power in case of a power failure.

2.7.2 *Short Questions*

These questions should be solved using only a simple, nonprogrammable, nongraphical calculator combined with pen and paper.

(31) Consider that you have been assigned the task of designing an automation system for a door to monitor who is present at the door and allow the occupant of the house to open the door if necessary. List what sensors, actuators, and other devices would be needed to accomplish this task.
(32) Consider the task of monitoring the temperature of a glass furnace. What considerations should you take into account? Which sensors would you use?
(33) Consider the task of pumping a mixture of sand, water, oil, air, and various particles. What would you need to consider when designing the pump? What kind of sensors would you consider? Do you think you can achieve highly accurate results?
(34) Consider the data shown in Table 2.2. Create the valve characterisation curve using the data. Determine what type of valve this is and how reproducible the values are. Are there any static or dynamic nonlinearities present? How can you determine this?
(35) Consider the sensor data shown in Table 2.3. It is desired to determine if the sensor values are properly calibrated against the measured values. Determine if the calibration is correct. Are the values reliable?

Table 2.2 Data for creating the valve characterisation curve for Question 34

%open	Flow rate, $\dot{m}$ (kg/min)		
	Run 1	Run 2	Run 3
0	0	0	0
10	0.5	0.5	0.4
20	2.3	2.3	2.3
30	4	4	4
40	5.6	5.5	5.5
50	6.8	6.8	6.8
60	8	8	7.9
70	8.9	8.9	8.9
80	9.7	9.7	9.7
90	10.4	10.5	10.5
100	11	11	11
90	10.7	10.7	10.7
80	10	10	10
70	9.2	9.2	9.2
60	8.3	8.3	8.4
50	7.2	7.2	7.2
40	6	6	6
30	4.6	4.6	4.6
20	3	3	3
10	1.2	1.2	1.2
0	0	0	0

Table 2.3 Sensor calibration data for Question 35

Measured height (m)	Sensor value (m)	
	Run 1	Run 2
0.00	0.02	0.03
0.10	0.115	0.105
0.15	0.149	0.152
0.20	0.229	0.215
0.25	0.248	0.251
0.30	0.321	0.312
0.35	0.348	0.349
0.40	0.412	0.392
0.45	0.452	0.457
0.50	0.512	0.493

Chapter 3
Mathematical Representation of a Process

In order to understand and provide useful information about a process, it is necessary to understand how different processes and systems can be represented. In practice, there exist two main types of representations: **mathematical** and **schematic**. A mathematical representation focuses on providing an abstract description of the process that provides information about the process. A good mathematical representation of the system can provide a deep understanding of how the system works and how it will behave in the future. On the other hand, a schematic representation focuses on the relationships between the different components and how they relate to each other. It is primarily a visual approach that allows the actual process to be represented on a piece of paper.

Common mathematical representations include **state-space models**, **transfer-function models**, and **automata**.

3.1 Laplace and *z*-Transforms

Before considering the mathematical models themselves, it is helpful to review two very common transformations between the time and frequency domains: the Laplace and *z*-transforms. If the time domain is continuous, then the **Laplace transform is** used, while if the time domain is discrete, then the ***z*-transform** is used.

3.1.1 Laplace Transform

The **Laplace transform** converts a function from the time domain into the Laplace (frequency) domain. This transform allows for a simple algebraic solution

Y. A. W. Shardt, *Automation Engineering*,
https://doi.org/10.1007/978-3-031-92394-4_3

of complex differential equations. Due to this feature, Laplace transforms are widely used in automation engineering to understand process behaviour and obtain solutions to various control problems.

The Laplace transform is defined as[1]

$$F(s) = \int_{0^-}^{\infty} f(t)e^{-st}dt \tag{3.1}$$

where F is the Laplace-transformed function, f the original function, and s the Laplace variable. Conventionally, the Laplace transform is denoted by a script L, $\mathcal{L}$, that is, $F(s)=\mathcal{L}(f(t))$. Converting from the Laplace domain to the time domain is conventionally shown using the inverse of the Laplace transform, $\mathcal{L}^{-1}$. Table 3.1 presents a summary of the most common Laplace transforms. The following are some useful properties of the Laplace transform:

(1) **Superposition**: $\mathcal{L}(f + g) = \mathcal{L}(f) + \mathcal{L}(g)$.

(2) **Homogeneity**: $\mathcal{L}(\alpha f) = \alpha\mathcal{L}(f), \alpha \in \mathbb{R}$.

(3) **Convolution**: $\mathcal{L}(f * g) = \mathcal{L}\left(\int_{0^-}^{t} f(\tau)g(t - \tau)d\tau\right) = F(s)G(s)$, where $*$ is the convolution operator.

(4) **Shifting Properties**: The following rules can be useful in solving problems involving Laplace transforms (where δ is the Dirac delta function, u the step response, and $a \geq 0$):

$$\mathcal{L}(f(t - a)u(t - a)) = e^{-as}F(s) \tag{3.2}$$

$$\mathcal{L}(g(t)u(t - a)) = e^{-as}\mathcal{L}(g(t + a)) \tag{3.3}$$

$$\mathcal{L}(\delta(t - a)) = e^{-as} \tag{3.4}$$

$$\mathcal{L}(f(t)\delta(t - a)) = f(a)e^{-as} \tag{3.5}$$

(5) **Final-Value Theorem**: Assuming that the poles (roots of the denominator) of $sF(s)$ lie in the left-hand plane,[2] then

$$\lim_{t\to\infty} f(t) = \lim_{s\to 0} sF(s) \tag{3.6}$$

[1]The notation 0^- essentially represents the left limit of the signal or the pre-initial condition. In most cases, this will be equivalent to the value at 0. However, for functions with discontinuities at 0, this will not be true. Similarly, 0^+ represents the right limit or the postinitial conditions. Additional information about this issue can be found in Lundberg et al. (2007).

[2]Equivalently, the real component of all the poles must be less than 0 for the system to be stable.

(6) **Initial-Value Theorem**: The value in the time domain at the starting point $t = 0^+$ is given by

$$\lim_{t \to 0^+} f(t) = \lim_{s \to \infty} sF(s) \tag{3.7}$$

Often, when we are given function in the Laplace domain, it may be necessary to convert it into the time domain. This can be performed by using the inverse Laplace transform, $\mathcal{L}^{-1}$, such that $\mathcal{L}^{-1}(\mathcal{L}(f(t)) = f(t)$. In most automation-engineering problems, the general case reduces to finding the time-domain function corresponding to some rational function of s. In such cases, partial fractioning (see Appendix I for the details) is required to break up the original fraction into its constituent parts so that the known functions given in Table 3.2 can be used.

Table 3.1 Table of common Laplace transforms

Case	Time domain $f(t)$	Laplace domain $F(s)$
Dirac delta or impulse function, δ	$\delta(t-a) = \begin{cases} 0 & t \neq a \\ \infty & t = a \end{cases}$	e^{-as}
Unit step function, u	$u(t) = \begin{cases} 0 & t \leq 0 \\ 1 & t > 0 \end{cases}$	s^{-1}
Polynomials	$\frac{t^{n-1}}{(n-1)!}$	$\frac{1}{s^n}$
Exponential	e^{-at}	$(s+a)^{-1}$
Cosine	$e^{-at}\cos(\omega t)$	$\frac{s+a}{(s+a)^2+\omega^2}$
Sine	$e^{-at}\sin(\omega t)$	$\frac{\omega}{(s+a)^2+\omega^2}$
Derivative[3]	$\frac{d^n f(t)}{dt^n} = f^{(n)}(t)$	$s^n F(s) - \sum_{k=1}^{n} s^{k-1} f^{(n-k)}(0^-)$
Integration	$\int_{0^-}^{\infty} f(t)dt$	$\frac{1}{s}F(s)$
Time shift (or time delay)	$f(t-a)u(t-a)$	$e^{-as}F(s)$

[3] If it is assumed that the system is initial at steady-state and using deviational variables, then all derivatives will be zero and this equation will reduce to the first term $s^n F(s)$.

Table 3.2 Table of useful inverse Laplace relationships

$\mathcal{L}(f(t))$	$f(t)$
$\frac{A}{Cs+D}$	$\frac{A}{C}\mathrm{e}^{\frac{-D}{C}t}$
$\frac{A}{(Cs+D)^n}$	$\frac{A}{(n-1)!C^n}t^{n-1}\mathrm{e}^{\frac{-D}{C}t}$
$\frac{Cs+E}{\alpha s^2+\beta s+\gamma}$ (irreducible quadratic)	$\left(\frac{E-\frac{C\beta}{2\alpha}}{\sqrt{\alpha\rho}}\right)\mathrm{e}^{\frac{-\beta}{2\alpha}t}\sin\left(t\sqrt{\frac{\rho}{\alpha}}\right)+\left(\frac{C}{\alpha}\right)\mathrm{e}^{\frac{-\beta}{2\alpha}t}\cos\left(t\sqrt{\frac{\rho}{\alpha}}\right)$ with $\rho=\gamma-\frac{\beta^2}{4\alpha}\geq 0$
$\frac{Cs+D}{\left(\alpha s^2+\beta s+\gamma\right)^n}$ (irreducible quadratic)	$\left(\frac{C}{\alpha^n(n-2)!}\right)\left[\left(t^{n-2}\mathrm{e}^{\frac{-\beta}{2\alpha}t}\right)\otimes\left(\mathrm{e}^{\frac{-\beta}{2\alpha}t}\cos\left(t\sqrt{\frac{\rho}{\alpha}}\right)\right)^{\otimes n}\right]+$ $\left(\frac{2\alpha D-C\beta}{2\alpha^{n+1}}\left(\frac{\alpha}{\rho}\right)^{n/2}\right)\left(\mathrm{e}^{\frac{-\beta}{2\alpha}t}\sin\left(t\sqrt{\frac{\rho}{\alpha}}\right)\right)^{\otimes n}$ with $n\neq 1$, $\rho=\gamma-\frac{\beta^2}{4\alpha}\geq 0$, $\otimes$ is convolution, and $(f)^{\otimes n}$ is defined as the convolution of f, n times, i.e. $\underbrace{f\otimes f\otimes f\cdots f}_{n\text{ times}}$

Example 3.1: Laplace Transform

Compute the Laplace transform for the function

$$y_t = t^5 + \mathrm{e}^{-5t}\cos(7t) \tag{3.8}$$

Solution

Since the Laplace transform is linear, we can determine the Laplace forms of each part separately and then combine them together. From Table 3.1, we see that

$$\mathcal{L}\left(\frac{t^{n-1}}{(n-1)!}\right) = \frac{1}{s^n} \tag{3.9}$$

Setting $n - 1 = 5$, which is the exponent of t^5 and noting that we will need to multiply both sides by the factorial $(n - 1)!$, will give a Laplace form of

$$\mathcal{L}\left(t^5\right) = \frac{5!}{s^6} \tag{3.10}$$

Similarly, for the second term, from Table 3.1, we have that

$$\mathcal{L}\left(\mathrm{e}^{-at}\cos(\omega t)\right) = \frac{s+a}{(s+a)^2+\omega^2} \tag{3.11}$$

Comparing with the form that we have, we see that $a = 5$ and $\omega = 7$. Thus, the Laplace transform is

$$\mathcal{L}\left(e^{-5t}\cos(7t)\right) = \frac{s+5}{(s+5)^2+7^2} \tag{3.12}$$

Combining the two parts together gives

$$\mathcal{L}\left(t^5 + e^{-5t}\cos(7t)\right) = \frac{5!}{s^6} + \frac{s+5}{(s+5)^2+7^2} \tag{3.13}$$

which is the Laplace transform of y_t.

Example 3.2: Inverse Laplace Transform
Compute the time-domain representation of the following partial-fractioned Laplace function

$$Y(s) = \frac{5}{10s+1} + \frac{8}{(2s+1)^5} \tag{3.14}$$

Solution

The solution will be found by treating each fraction separately and using the information in Table 3.2. For the first term, the general form of the transform can be found from Table 3.2 as

$$\mathcal{L}^{-1}\left(\frac{A}{Cs+D}\right) = \frac{A}{C}e^{\frac{-D}{C}t} \tag{3.15}$$

Comparing the general form with our first term, we see that $A = 5$, $C = 10$, and $D = 1$, which implies that the inverse Laplace form will be

$$\mathcal{L}^{-1}\left(\frac{5}{10s+1}\right) = \frac{5}{10}e^{\frac{-1}{10}t} = 0.5e^{-0.1t} \tag{3.16}$$

For the second term, the general form can be written as

$$\mathcal{L}^{-1}\left(\frac{A}{(Cs+D)^n}\right) = \frac{A}{(n-1)!C^n}t^{n-1}e^{\frac{-D}{C}t} \tag{3.17}$$

Comparing the general form with our second term, we see that $A = 8$, $C = 2$, $D = 1$, and $n = 5$, which implies that the inverse Laplace form will be

$$\mathcal{L}^{-1}\left(\frac{8}{(2s+1)^5}\right) = \frac{8}{(5-1)!2^5}t^{5-1}e^{\frac{-1}{2}t} = \frac{1}{96}t^4e^{-0.5t} \tag{3.18}$$

Combining the two terms together gives

$$\mathcal{L}^{-1}\left(\frac{5}{10s+1} + \frac{8}{(2s+1)^5}\right) = 0.5e^{-0.1t} + \frac{1}{96}t^4e^{-0.5t} \tag{3.19}$$

This gives us the time-domain representation of the original Laplace function.

3.1.2 z-Transform

The **z-transform** converts a discrete function from the time domain into the z (frequency) domain. This transform allows for a simple algebraic solution of complex difference equations. Due to this feature, z-transforms are widely used in automation engineering to understand process behaviour and obtain solutions to various control problems.

The z-transform is defined as

$$F(z) = \sum_{n=0}^{\infty} f_n z^{-n} \tag{3.20}$$

where F is the transformed function and f the original discrete function. Conventionally, the z-transform is denoted by a script $\mathcal{Z}$. Table 3.3 presents a summary of the most common z-transforms. The following are some useful properties:

(1) **Superposition**: $\mathcal{Z}(f + g) = \mathcal{Z}(f) + \mathcal{Z}(g)$.
(2) **Homogeneity**: $\mathcal{Z}(\alpha f) = \alpha \mathcal{Z}(f), \alpha \in \mathbb{R}$.
(3) **Final-Value Theorem**: Assuming that the poles (roots of the denominator) of $(z - 1)F(z)$ lie inside the unit circle, then[4]

$$\lim_{k \to \infty} f_k = \lim_{z \to 1} (z - 1)F(z) \tag{3.21}$$

(4) **Initial-Value Theorem**: The value in the time domain at the starting point $k=0$ is given by

$$\lim_{k \to 0} f_k = \lim_{z \to \infty} zF(z) \tag{3.22}$$

As can be seen from Table 3.3, many of the forms involve z^{-1}. For this reason, in automation engineering, it is common to treat z^{-1} as the variable, which is called the **backshift operator**. Converting between the two representations is rather easy as it involves multiplying by the highest power present in the equation.

The inverse operation of finding the time-domain representation given the z-transform is performed using the inverse z-transform. Since most automation-engineering examples consider rational functions of z, this will require partial fractioning in order to split the rational function into its constituent parts (see Appendix I for details). Once the constituent parts have been obtained, we can then use Table 3.4 to find the corresponding time-domain representation. Instead of partial-fractioning, it may be possible to perform long division to obtain the individual values. However, this can be quite long and difficult to do.

---

[4] The roots are expressed in terms of z. Equivalently, the system is stable.

Table 3.3 Table of common z-transforms (T_s is the sampling time)

Case	Continuous-time domain $f(t)$	Discrete-time domain ($f(kT_s)$)	z-domain $F(s)$
Dirac delta, δ[5]	$\delta(t-a)=\begin{cases}0 & t\neq a\\ \infty & t=a\end{cases}$	$\delta_{ka}=\begin{cases}0 & k\neq a\\ 1 & k=a\end{cases}$	z^{-a}
Step function, u	$u(t)=\begin{cases}0 & t\leq 0\\ 1 & t>0\end{cases}$	$u_k=\begin{cases}0 & k\leq 0\\ 1 & k>0\end{cases}$	$\frac{1}{1-z^{-1}}$
Exponential	e^{at}	e^{akT_s}	$\frac{1}{1-e^{aT_s}z^{-1}}$
Exponential + cosine	$e^{at}\cos(\omega t)$	$e^{akT_s}\cos(\omega kT_s)$	$\frac{1-e^{aT_s}\cos(\omega T_s)z^{-1}}{1-2e^{aT_s}\cos(\omega T_s)z^{-1}+e^{2aT_s}z^{-2}}$
Exponential + sine	$e^{at}\sin(\omega t)$	$e^{akT_s}\sin(\omega kT_s)$	$\frac{e^{aT_s}\sin(\omega T_s)z^{-1}}{1-2e^{aT_s}\cos(\omega T_s)z^{-1}+e^{2aT_s}z^{-2}}$
General power series		a^k	$\frac{1}{1-az^{-1}}$
First difference (derivative)	$\frac{df}{d}t$	f_k-f_{k-1}	$(1-z^{-1})F(z)$
Time shift (delay)	$f(t-a)u(t-a)$	$f_{k-a}u_{k-a}$	$z^{-a}F(z)$

Table 3.4 Useful inverse z-transform table (A and D can take complex values)

$\mathcal{Z}(y_k)$	y_k
$\frac{Az}{Cz+D}=\frac{A}{C+Dz^{-1}}$	$y_k=\frac{A}{C}\left(-\frac{D}{C}\right)^k$
$\frac{Az}{(Cz+D)^n}$	$y_k=\frac{A}{C^n}\left(\frac{\prod_{j=1}^{n-1}(k-j+1)}{\alpha^{n-1}(n-1)!}\right)\alpha^k, \alpha=-\frac{D}{C}$
$\frac{A}{(C+Dz^{-1})^n}, n\in\mathbb{Z}$	$y_k=\frac{A}{C^n}\begin{pmatrix}k+n-1\\ n-1\end{pmatrix}\alpha^k, \alpha=-\frac{D}{C}$

Example 3.3: z-Transform

Compute the z-transform of the following function

$$y_k=\begin{cases}0 & k<2\\ 5^{k-2}, & k\geq 2\end{cases} \tag{3.23}$$

Solution

From Table 3.3, we see that an exponential function a^k has the z-transform

$$\frac{1}{1-az^{-1}} \tag{3.24}$$

In our case, this implies that $a=5$. However, we can note that the values are delayed by 2 samples. Therefore, we will need to also use the delay formula from Table 3.3 to obtain the final result. Thus, the z-transform is

[5] In the discrete domain, the Dirac delta function is most often called the *Kronecker delta function*.

$$\mathcal{Z}(5^{k-2}) = \frac{z^{-2}}{1 - 5z^{-1}} \tag{3.25}$$

Example 3.4: Inverse z-Transform

Compute for the following frequency-domain function

$$\frac{2}{1 + 5z^{-1}} \tag{3.26}$$

the corresponding time-domain representation.

Solution

From Table 3.4, we can see that this represents the first case with $A=2$, $C=1$, and $D=5$. This implies that the time-domain representation is

$$\mathcal{Z}^{-1}\left(\frac{2}{1 + 5z^{-1}}\right) = \frac{A}{C}\left(-\frac{D}{C}\right)^k = \frac{2}{1}\left(-\frac{5}{1}\right)^k = 2(-5)^k \tag{3.27}$$

Finally, we can note that there is a relationship between the Laplace transform of the continuous-time domain and the z-transform of the discrete-time domain, namely,

$$z = e^{sT} \tag{3.28}$$

where T is the sampling time. This means that results obtained in the continuous domain will have a transformed representation in the discrete domain. Of note, the imaginary axis of the continuous domain is mapped onto the boundary of the unit circle.

3.2 Time- and Frequency-Based Models

Time- and frequency-based models are one of the most commonly encountered mathematical representations of a process. Time-based models focus on the behaviour of the system with respect to time, that is, how the system evolves or changes over time given specific inputs and states. However, since solutions can only often be obtained by integrating the complex differential equations, recourse is often made to frequency-based models where it can be easier to understand how the process will behave when the inputs change.

Before looking into the different types of time- and frequency-based models, it can be useful to consider some of the terms that can be used to classify the different types of models:

(1) **Linear** *versus* **nonlinear**: A model $f(t)$ is said to be **linear** with respect to t if the following two statements hold:

a. **Principle of Superposition**: $f(t_1+t_2)=f(t_1)+f(t_2)$, and
b. **Principle of Homogeneity**: $f(\alpha t_1)=\alpha f(t_1)$, $\alpha \in \mathbb{R}$.

If these two statements do not hold, then the model is said to be **nonlinear**. Linearity is a very useful property that allows for a system to be easily analysed using known, well-established theoretical methods. Nonlinear models can be linearised, so that in a given region they are well described by the linear equivalent model. A model can be linear with respect to one variable, but not another.

(2) **Time-invariant** *versus* **time-variant**: A model is said to be time-invariant if the parameters in the model are constant with respect to time. In a time-varying system, the model parameters can depend on the time period.
(3) **Lumped parameter** *versus* **distributed parameter**: A model is said to be a lumped-parameter system if the model does not depend on the location, that is, there are no space derivatives present. A model that depends on the location, that is, it contains space derivatives, is called a distributed-parameter system. For example, if the temperature T is a function of the x- or y-direction, that is, we have a derivative of the form $\frac{\partial^2 T}{\partial x^2}$ in the model, then the model is a distributed-parameter system. If the temperature only depends on the time, then we have a lumped-parameter system. The analysis of distributed-parameter systems is often much more difficult than that of a lumped-parameter system. A distributed-parameter system can be reduced to a lumped-parameter model if it is assumed that the variables are homogenously distributed within the space, and hence, all the space derivatives are zero. An extension of this idea is to discretise the space into a set of regions that can be assumed to be modelled by a lumped-parameter equivalent and then solving the resulting system of equations.
(4) **Causal** *versus* **noncausal**: A system is said to be causal if the future values only depend on the current and past values. In a noncausal system, the future values depend on the current, past, and future values. A noncausal system is not physically realisable, since the future values will never be known.
(5) **System with memory (dynamic)** *versus* **system without memory (static)**: A system is said to have memory if the future output depends on both the past and present inputs and states. On the other hand, a system is said to be without memory or memoryless if the future output only depends on the current input.

3.2.1 Time- and Frequency-Domain Representations

In automation engineering, there are two common representations of a system: **state-space** and **transfer-function** models.

The **state-space model** focuses on the relationship between states, inputs, and outputs in the time domain. The general state-space model is given as

$$\begin{aligned}\frac{\mathrm{d}\vec{x}}{\mathrm{d}t} &= \vec{f}(\vec{x}, \vec{u}) \\ \vec{y} &= \vec{g}(\vec{x}, \vec{u})\end{aligned} \tag{3.29}$$

where x is the **state** variable, u is the **input** variable, t is the time, y is an **output**, f and g are some functions, and an arrow above denotes a vector. The state variable is a variable that describes the current location of the system from which the system's future behaviour can be determined. A state variable will often appear in some equation as a derivative with respect to time. The **input** variable, u, is a variable that describes the properties of a stream entering a system. Traditionally, the number of inputs is denoted by m, the number of states by n, and the number of outputs by p. A system where $p = m = 1$ is said to be **univariate** or **single-input, single-output (SISO)**. If p and m are greater than 1, then the system is said to be **multivariate** or **multi-input, multi-output (MIMO)**. A system is said to be **multi-input, single-output** (MISO) if $p = 1$ and $m > 1$.

The general state-space model is often reduced to a linear form

$$\begin{aligned}\frac{\mathrm{d}\vec{x}}{\mathrm{d}t} &= \mathcal{A}\vec{x} + \mathcal{B}\vec{u} \\ \vec{y} &= \mathcal{C}\vec{x} + \mathcal{D}\vec{u}\end{aligned} \tag{3.30}$$

where $\mathcal{A}$ is the $n \times n$ state matrix, $\mathcal{B}$ is the $n \times m$ input matrix, $\mathcal{C}$ is the $p \times n$ output matrix, and $\mathcal{D}$ is the $p \times m$ feed-through matrix.

On the other hand, the **transfer-function representation** focuses solely on the relationship between the inputs and outputs in the Laplace domain and allows easier analysis of the system than for a state-space model. The general transfer-function representation can be written as

$$\vec{Y}(s) = \mathcal{G}(s)\vec{U}(s) \tag{3.31}$$

where Y is the output in the Laplace domain, U the input in the Laplace domain, and $\mathcal{G}$ a matrix containing the transfer function representation of the model, that is

$$\vec{Y}(s) = \begin{bmatrix} Y_1(s) \\ \vdots \\ Y_p(s) \end{bmatrix}, \vec{U}(s) = \begin{bmatrix} U_1(s) \\ \vdots \\ U_m(s) \end{bmatrix}, \mathcal{G}(s) = \begin{bmatrix} G_{11}(s) & \cdots & G_{1m}(s) \\ \vdots & & \vdots \\ G_{p1}(s) & \cdots & G_{pm}(s) \end{bmatrix} \tag{3.32}$$

It is often assumed that each transfer function can be written as

$$G(s) = \frac{N(s)}{D(s)} e^{-\theta s} \tag{3.33}$$

where N and D are polynomials of s and θ is the **deadtime** or **time delay** in the system. Time delay arises in real systems due to transport phenomena, measurement delays, and approximations introduced when linearising a system. The **order of a transfer function** is equal to the highest power of the D-polynomial.

Given the simple form of a transfer-function representation, much of the analysis in control is performed using it.

3.2.2 Converting Between Representations

The state-space model can be converted to a transfer-function model, by applying the Laplace transform to Eq. (3.30) and re-arranging to give[6]

$$\begin{aligned} s\vec{X} &= \mathcal{A}\vec{X} + \mathcal{B}\vec{U} \\ \vec{Y} &= \mathcal{C}\vec{X} + \mathcal{D}\vec{U} \end{aligned} \tag{3.34}$$

[7]

$$\begin{aligned} \vec{X} &= (s\mathcal{I} - \mathcal{A})^{-1}\mathcal{B}\vec{U} \\ \vec{Y} &= \mathcal{C}\vec{X} + \mathcal{D}\vec{U} \end{aligned} \tag{3.35}$$

$$\boxed{\mathcal{G}(s) = \frac{\vec{Y}}{\vec{U}} = \mathcal{C}(s\mathcal{I} - \mathcal{A})^{-1}\mathcal{B} + \mathcal{D}} \tag{3.36}$$

where $\mathcal{I}$ is the $n \times n$ identity matrix and s is the Laplace transform variable.

Example 3.5: Numeric Example of Obtaining a Transfer Function
Consider the example

$$\begin{cases} 5\dfrac{\mathrm{d}x}{\mathrm{d}t} + 3x = u \\ y = x \end{cases} \tag{3.37}$$

Taking the Laplace transform of Eq. (3.37) gives

[6] In automation engineering, we often deal with deviation variables that are defined as $\tilde{x} = x - x_{ss}$, where x_{ss} is some steady-state value. Since we assume that the process is initially at steady state, this means that the initial conditions will always be zero.

[7] The inverse exists since the $\mathcal{A}$-matrix only contains numeric entries.

$$\begin{cases} 5sX(s) + 3X(s) = U(s) \\ Y(s) = X(s) \end{cases} \tag{3.38}$$

Substituting Y for X in the first equation of Eq. (3.38) gives

$$5sY(s) + 3Y(s) = U(s) \tag{3.39}$$

Solving for Y/U gives

$$\begin{aligned} (5s+3)Y(s) &= U(s) \\ \frac{Y(s)}{U(s)} &= \frac{1}{5s+3} \end{aligned} \tag{3.40}$$

Therefore, the transfer function is

$$\boxed{G(s) = \frac{Y(s)}{U(s)} = \frac{1}{5s+3}} \tag{3.41}$$

Example 3.6: General Univariate Case

Consider the following Nth-order ordinary differential equation

$$\begin{cases} \frac{d^n\tilde{x}}{dt^n} + a_{n-1}\frac{d^{n-1}\tilde{x}}{dt^{n-1}} + \cdots + a_0\tilde{x} = b_{n-1}\frac{d^{n-1}\tilde{u}}{dt^{n-1}} + \cdots + b_0\tilde{u} \\ \tilde{y} = \tilde{x} \end{cases} \tag{3.42}$$

such that at $t=0$, all the derivatives are equal to zero and $\tilde{x}=\tilde{y}=0$. Taking the Laplace transform of Eq. (3.42) and simplifying gives

$$\begin{gathered} \begin{cases} s^nX(s) + a_{n-1}s^{n-1}X(s) + \cdots + a_0X(s) = b_{n-1}s^{n-1}U(s) + \cdots + b_0U(s) \\ Y(s) = X(s) \end{cases} \\ s^nY(s) + a_{n-1}s^{n-1}Y(s) + \cdots + a_0Y(s) = b_{n-1}s^{n-1}U(s) + \cdots + b_0U(s) \end{gathered} \tag{3.43}$$

Collecting like terms and re-arranging gives

$$\begin{aligned} (s^n + a_{n-1}s^{n-1} + \cdots + a_0)Y(s) &= (b_{n-1}s^{n-1} + \cdots + b_0)U(s) \\ G(s) = \frac{Y(s)}{U(s)} &= \frac{(b_{n-1}s^{n-1} + \cdots + b_0)}{(s^n + a_{n-1}s^{n-1} + \cdots + a_0)} \end{aligned} \tag{3.44}$$

Example 3.7: Multivariate Example
Consider the following differential equation

$$\begin{cases} \dfrac{\mathrm{d}\tilde{x}}{\mathrm{d}t} = -a\tilde{x} + b_1\tilde{u}_1 + b_2\tilde{u}_2 \\ \tilde{y} = \tilde{x} \end{cases} \tag{3.45}$$

and determine the transfer functions for the system. Note that there are two inputs ($m=2$) and one output ($p=1$).

Solution

$$\begin{aligned} &\begin{cases} sX(s) = -aX(s) + b_1U_1(s) + b_2U_2(s) \\ Y(s) = X(s) \end{cases} \\ &sY(s) + aY(s) = b_1U_1(s) + b_2U_2(s) \\ &(s+a)Y(s) = b_1U_1(s) + b_2U_2(s) \\ &Y(s) = \frac{b_1}{s+a}U_1(s) + \frac{b_2}{s+a}U_2(s) \end{aligned} \tag{3.46}$$

Rewriting this into matrix form gives

$$Y(s) = \begin{bmatrix} \frac{b_1}{s+a} & \frac{b_2}{s+a} \end{bmatrix} \begin{bmatrix} U_1(s) \\ U_2(s) \end{bmatrix} \tag{3.47}$$

Note that this can be written separately, since linear functions are being considered. As well, note that the principle of superposition holds, so that we can study each transfer function separately and then combine the results together.

The reverse operation of converting a transfer function into a state-space representation does not yield a unique solution or **realisation**. Consider the following transfer function, where $p<n$:

$$G(s) = \frac{\beta_p s^p + \beta_{p-1}s^{p-1} + \cdots + \beta_1 s + \beta_0}{s^n + \alpha_{n-1}s^{n-1} + \cdots + \alpha_1 s + \alpha_0} \tag{3.48}$$

then the **controllable canonical realisation** is

$$\mathcal{A}=\begin{bmatrix} -\alpha_{n-1} & -\alpha_{n-2} & \cdots & -\alpha_1 & -\alpha_0 \\ 1 & 0 & \cdots & 0 & 0 \\ 0 & 1 & \ddots & \vdots & \vdots \\ \vdots & \ddots & \ddots & 0 & \vdots \\ 0 & \cdots & 0 & 1 & 0 \end{bmatrix}_{n\times n} \quad \mathcal{B}=\begin{bmatrix} 1 \\ 0 \\ \vdots \\ \vdots \\ 0 \end{bmatrix}_{n\times 1} \tag{3.49}$$

$$\mathcal{C}=\begin{bmatrix} 0 & \cdots & 0 & \beta_p & \beta_{p-1} & \cdots & \beta_1 & \beta_0 \end{bmatrix}_{1\times n} \quad \mathcal{D}=0_{1\times 1}$$

while the **observable canonical realisation** is

$$\mathcal{A}=\begin{bmatrix} -\alpha_{n-1} & -\alpha_{n-2} & \cdots & -\alpha_1 & -\alpha_0 \\ 1 & 0 & \cdots & 0 & 0 \\ 0 & 1 & \ddots & \vdots & \vdots \\ \vdots & \ddots & \ddots & 0 & \vdots \\ 0 & \cdots & 0 & 1 & 0 \end{bmatrix}_{n\times n}^{T} \quad \mathcal{C}=\begin{bmatrix} 1 \\ 0 \\ \vdots \\ \vdots \\ 0 \end{bmatrix}_{n\times 1}^{T} \tag{3.50}$$

$$\mathcal{B}=\begin{bmatrix} 0 & \cdots & 0 & \beta_p & \beta_{p-1} & \cdots & \beta_1 & \beta_0 \end{bmatrix}_{1\times n}^{T} \quad \mathcal{D}=0_{1\times 1}$$

Note that for $\mathcal{C}$ in the controllable canonical realisation and for $\mathcal{B}$ in the observable canonical realisation, there will be $n-p-1$ zeroes at the beginning of the vector. The fact that the two forms are related by taking the transpose is not coincidental.

Example 3.8: Converting a Transfer Function into Its Controllable Canonical Realisation

Convert the transfer function

$$G(s)=\frac{s-2}{s^2-4s+1} \tag{3.51}$$

into its controllable canonical realisation.

Solution

First, we need to make sure that the transfer function is the form given by Eq. (3.48) and determine the values of the parameters. Since the equations have the same form, we note that $n=2$ (highest power in the denominator) and $p=1$ (highest power in the numerator), which implies that the transfer function satisfies the requirements. Thus, we can simply compare the parameters and obtain their values, that is, $\alpha_2 = 1$, $\alpha_1 = -4$, $\alpha_0 = 1$, $\beta_1 = 1$, and $\beta_0 = -2$. Thus, using Eq. (3.49), the controllable canonical realisation is

$$\begin{gathered} \mathcal{A} = \begin{bmatrix} 4 & -1 \\ 1 & 0 \end{bmatrix}, \ \mathcal{B} = \begin{bmatrix} 1 \\ 0 \end{bmatrix} \\ \mathcal{C} = \begin{bmatrix} 1 & -2 \end{bmatrix}, \ \mathcal{D} = 0_{1\times 1} \end{gathered} \tag{3.52}$$

Thus, we have defined the four matrices for one of the state-space realisations for the given transfer function.

3.2.3 Discrete-Time Models

In the discrete domain, the form and types of available models are similar. The linearised state-space representation can be written as

$$\begin{aligned} \vec{x}_{t+1} &= \mathcal{A}_d \vec{x}_t + \mathcal{B}_d \vec{u}_t \\ \vec{y}_t &= \mathcal{C}_d \vec{x}_t + \mathcal{D}_d \vec{u}_t \end{aligned} \tag{3.53}$$

The subscript d in Eq. (3.53) can be dropped if it is clear that a discrete state-space representation is being considered. The discrete transfer function is the same as the continuous version except that all s's are replaced by either z or z^{-1}. To convert between the discrete state-space and transfer function representations, Eq. (3.36), can be used, *mutatis mutandis*.

With discrete transfer functions, it is common to explicitly include the unmeasured disturbance, denoted by e_t, in the final model. In such cases, the unmeasured disturbance signal is assumed to be a stochastic (random) Gaussian, white-noise signal. A Gaussian, white-noise signal implies that the values of this signal are normally distributed and independent of past or future values. The general discrete transfer function can be written as

$$y_t = G_p\left(z^{-1}, \vec{\theta}\right) u_t + G_l\left(z^{-1}, \vec{\theta}\right) e_t \tag{3.54}$$

where G_p is the process transfer function, θ the parameters, and G_l the disturbance transfer function. Since in most applications, it is assumed that the transfer functions are rational functions of z^{-1}, the most common discrete transfer function equation can be written in a form called the **prediction-error model**, which has the following form:

$$A(z^{-1}) y_t = \frac{B(z^{-1})}{F(z^{-1})} u_{t-k} + \frac{C(z^{-1})}{D(z^{-1})} e_t \tag{3.55}$$

where $A(z^{-1})$, $C(z^{-1})$, $D(z^{-1})$, and $F(z^{-1})$ are polynomials in z^{-1} of the form

$$1+\sum_{i=1}^{n_a}\theta_i z^{-i} \tag{3.56}$$

n_a is the order of the polynomial and θ_i are the parameters, $B(z^{-1})$, is a polynomial in z^{-1} of the form

$$\sum_{i=1}^{n_b}\theta_i z^{-i} \tag{3.57}$$

n_b is the order of the polynomial, and k is the time delay in the system. In general, it is very rare for this model to be used directly. Instead, any of the following simplifications may be used:

(1) **Box-Jenkins Model**: In this model, the $A(z^{-1})$ polynomial is ignored. Thus, this model is given as

$$y_t=\frac{B(z^{-1})}{F(z^{-1})}u_{t-k}+\frac{C(z^{-1})}{D(z^{-1})}\mathrm{e}_t \tag{3.58}$$

In practice, this method is sufficient to fit an accurate model of the system.

(2) **Autoregressive Moving-Average Exogenous Model (ARMAX)**: In this model, the $D(z^{-1})$ and $F(z^{-1})$ polynomials are ignored, which gives

$$A(z^{-1})y_t=B(z^{-1})u_{t-k}+C(z^{-1})\mathrm{e}_t \tag{3.59}$$

This model assumes that the denominator for both the input and the error is the same. A further simplification is to ignore the $C(z^{-1})$ term. This gives an **autoregressive exogenous model (ARX)**. This model has the form given by:

$$A(z^{-1})y_t=B(z^{-1})u_{t-k}+\mathrm{e}_t \tag{3.60}$$

(3) **Output-ErrorModel (OE)**: In this model, only the model for the input is fit to the data. The error terms are ignored. Thus, the model is given as

$$y_t=\frac{B(z^{-1})}{F(z^{-1})}u_{t-k}+\mathrm{e}_t \tag{3.61}$$

3.2.4 Converting Between Discrete and Continuous Models

It is possible to convert between continuous and discrete forms of a model under certain assumption regarding the discretisation. The most common assumption is that the input stays constant between sampling instances, that is, a zero-order hold is used to convert from the continuous (analogue) to discrete (digital) domains. In such case, it can be shown that the continuous state-space representation can be converted into the discrete time using the following formulae:

$$\begin{aligned}\mathcal{A}_d &= \mathrm{e}^{\mathcal{A}\tau_s}\\ \mathcal{B}_d &= \left(\int\limits_{\tau=0}^{\tau=\tau_s} \mathrm{e}^{\mathcal{A}\tau}\,d\tau\right)\mathcal{B}\\ \mathcal{C}_d &= \mathcal{C}\\ \mathcal{D}_d &= \mathcal{D}\end{aligned} \tag{3.62}$$

where τ_s is the sampling time and $\mathrm{e}^{\mathcal{A}\tau}$ is the matrix exponential function. The corresponding transfer function can then be obtained using Eq. (3.36). For a simple, first-order, continuous transfer function given as

$$Y(s) = \frac{K}{\tau_p s + 1} U(s) \tag{3.63}$$

the discrete, transfer function is given as

$$y_t = \frac{K\left(1 - \mathrm{e}^{-\frac{\tau_p}{\tau_s}}\right)}{1 - \left(1 - \mathrm{e}^{-\frac{\tau_p}{\tau_s}}\right)z^{-1}} u_t \tag{3.64}$$

If a system has time delay, then it can be converted using the following formula

$$k = \left\lfloor \frac{\theta}{\tau_s} \right\rfloor \tag{3.65}$$

where θ is the continuous time delay and $\lfloor\cdot\rfloor$ is an appropriately selected rounding function that converts to an integer value.

The general rule of thumb for selecting the sampling time is that

$$\tau_s = (0.1 \text{ to } 0.2)\tau_{p,\,\min} \tag{3.66}$$

where $\tau_{p,\,\min}$ is the smallest time constant present in the system.

3.2.5 Impulse-Response Model

In the discrete domain, the **infinite-impulse-response** model is defined as

$$y_t = \sum_{i=0}^{\infty} h_i z^{-i} u_t = \sum_{i=0}^{\infty} h_i u_{t-i} \tag{3.67}$$

where h are the impulse coefficients. The values of h can be obtained by either performing long division on the transfer function or partial fractioning the result to obtain the impulse coefficients. Since the values of h can often quickly taper off, it is possible to convert the infinite-impulse-response model into a **finite-impulse-response** (FIR) model, that is,

$$y_t = \sum_{i=0}^{n} h_i z^{-i} u_t = \sum_{i=0}^{n} h_i u_{t-i} \tag{3.68}$$

where n is an integer representing the number of terms selected for the model.

3.2.6 Compact State-Space Representation

In many theoretical applications, the state-space representation is often written in a compact manner that resembles a transfer-function formulism but using matrices. For the standard state-space representation given by Eq. (3.30), the **compact** or block **state-space representation** can be written as

$$G = \left[\begin{array}{c|c} \mathcal{A} & \mathcal{B} \\ \hline \mathcal{C} & \mathcal{D} \end{array}\right] \tag{3.69}$$

Using this representation, it is possible to easily partition the matrices and show this in a compact manner, for example,

$$G = \left[\begin{array}{cc|c} \mathcal{A}_{11} & \mathcal{A}_{12} & \mathcal{B}_1 \\ \mathcal{A}_{21} & \mathcal{A}_{22} & \mathcal{B}_2 \\ \hline \mathcal{C}_1 & \mathcal{C}_2 & \mathcal{D} \end{array}\right] \tag{3.70}$$

where the states have been split into four groups and the relationships between the different groups can easily be shown.

When using the compact representation, various short cuts can be used when combining two models. When adding two compact representations, that is the transfer functions are in parallel,

$$G_1 + G_2 = \left[\begin{array}{cc|c} \mathcal{A}_1 & 0 & \mathcal{B}_1 \\ 0 & \mathcal{A}_2 & \mathcal{B}_2 \\ \hline \mathcal{C}_1 & \mathcal{C}_2 & \mathcal{D}_1 + \mathcal{D}_2 \end{array}\right] \tag{3.71}$$

When multiplying two representations, that is the transfer functions are in series,

$$G_1G_2 = \left[\begin{array}{cc|c} \mathcal{A}_1 & \mathcal{B}_1\mathcal{C}_2 & \mathcal{B}_1\mathcal{D}_2 \\ 0 & \mathcal{A}_2 & \mathcal{B}_2 \\ \hline \mathcal{C}_1 & \mathcal{D}_1\mathcal{C}_2 & \mathcal{D}_1\mathcal{D}_2 \end{array}\right] \tag{3.72}$$

The inverse of a compact state-space representation can be written as

$$G^{-1} = \left[\begin{array}{c|c} \mathcal{A} - \mathcal{B}\mathcal{D}^{-1}\mathcal{C} & -\mathcal{B}\mathcal{D}^{-1} \\ \hline \mathcal{D}^{-1}\mathcal{C} & \mathcal{D}^{-1} \end{array}\right] \tag{3.73}$$

provided that $\mathcal{D}$ is invertible. Finally, the transpose of this representation can be written as

$$G^T = \left[\begin{array}{c|c} \mathcal{A}^T & \mathcal{C}^T \\ \hline \mathcal{B}^T & \mathcal{D}^T \end{array}\right] \tag{3.74}$$

A common operation with state-space models is to transform the order or meaning of the states. In such cases, the compact representation can be derived as follows. Assume that

$$\begin{aligned} \tilde{x} &= \mathcal{T}x \\ \tilde{u} &= \mathcal{S}u \\ \tilde{y} &= \mathcal{R}y \end{aligned} \tag{3.75}$$

where $\mathcal{T}$, $\mathcal{R}$, and $\mathcal{S}$ are appropriately sized nonsingular (invertible) matrices. In this case, the transformed compact representation can be written as

$$\tilde{G} = \left[\begin{array}{c|c} \mathcal{T}\mathcal{A}\mathcal{T}^{-1} & \mathcal{T}\mathcal{B}\mathcal{S}^{-1} \\ \hline \mathcal{R}\mathcal{C}\mathcal{T}^{-1} & \mathcal{R}\mathcal{D}\mathcal{S}^{-1} \end{array}\right] \tag{3.76}$$

3.3 Process Analysis

Having considered the different models, it would be useful to understand what information can be easily extracted from such models that would allow us to understand how the process behaves. Since it is easier to extract information from transfer functions, the focus will be on analysing them. It will be assumed that the transfer function can be written in the continuous time domain as

$$G(s) = \frac{N(s)}{D(s)}\mathrm{e}^{-\theta s} \tag{3.77}$$

or in the discrete time domain as

$$G(z) = \frac{N(z)}{D(z)} z^{-k} \tag{3.78}$$

The **order of a polynomial**, denoted by n, is defined as the highest power present in a given polynomial, for example, x^4+x^2 is a fourth-order polynomial, since the highest power is 4. The **order of a transfer function** is equal to the order of the denominator polynomial, D. A transfer function is said to be **proper** if the order of the numerator is less than or equal to the order of the denominator, that is, $n_N \leq n_D$. A transfer function is said to be **strictly proper** if the order of the numerator is less than the order of the denominator, that is, $n_N < n_D$. For most physical processes, a transfer function will be strictly proper.

A process is said to be **causal** if the time delay, θ (or k) is nonnegative; otherwise, the process is **noncausal**. In the discrete domain, causality is ensured by a proper transfer function.

The **poles** of a transfer function are defined as the roots of the denominator, D, that is, those values of s (or z) such that $D = 0$. The **zeros** of a transfer function are defined as the roots of the numerator, N.

A process is said to be at **steady state** if all derivatives are equal to zero, that is,

$$\frac{d\vec{x}}{dt} = 0 \tag{3.79}$$

If Eq. (3.79) does not hold, then the process is said to be operating in **transient** mode. It should be noted that small deviations from zero do not necessarily mean that the process is now in a transient mode. In practice, it may not be possible to achieve an exact steady state, where all the derivatives are exactly equal to zero, both due to continual small fluctuations and measurement imprecisions. In such cases, it is common to define the **settling time**, t_s, of a process as the time it takes for the process to reach and stay within an envelope centred on the new steady-state value and has bounds that are 2.5% of the change in the process variable. Figure 3.1 shows how the settling time is computed.

For a process modelled by a continuous transfer function, there are three parameters of interest: **gain, time constant,** and **time delay** (**deadtime** or **process delay**). The **gain**, K, of the process represents the steady-state behaviour of the system for a unit change in the input. It is defined as

$$K = \frac{y(t \to \infty)}{u(t \to \infty)} \tag{3.80}$$

where it is assumed that the input is bounded as $t \to \infty$. The gain is the same irrespective of the bounded input used. Using the final-value theorem for a general transfer function and a step change in the input, it can be seen that

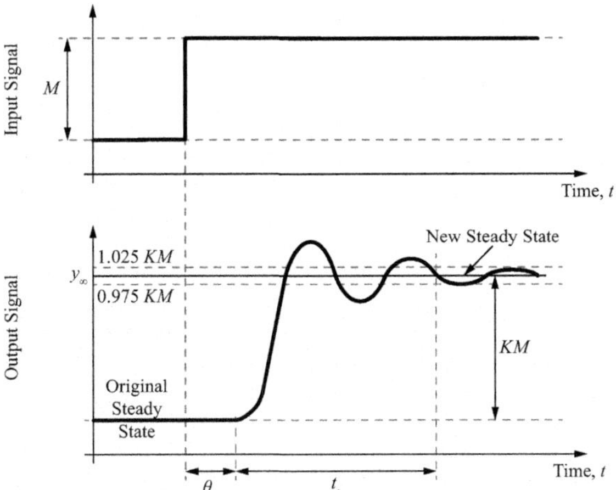

Fig. 3.1 Determining the settling time

$$
\begin{aligned}
K &= \lim_{s\to 0} sY(s) = \lim_{s\to 0} \frac{sG(s)}{s} = \lim_{s\to 0} G(s) \\
&= \lim_{s\to 0} \frac{\left(b_{n-1}s^{n-1} + \cdots + b_0\right)}{\left(s^n + a_{n-1}s^{n-1} + \cdots + a_0\right)} \mathrm{e}^{-\theta s} \\
&= \frac{b_0}{a_0}
\end{aligned}
\tag{3.81}
$$

This ratio b_0/a_0 is called the gain, K, of the process.

The **process time constant** τ represents the transient or dynamic component of the system, that is, how quickly the system responds to changes in the input and reaches a new steady-state value. The larger the time constant, the longer it takes to reach steady state. It can be obtained from the transfer function by factoring the denominator into the form

$$
\frac{\left(b_{n-1}s^{n-1} + \cdots + b_0\right)}{\left(s^n + a_{n-1}s^{n-1} + \cdots + a_0\right)} = \frac{\left(b_{n-1}s^{n-1} + \cdots + b_0\right)}{\prod_{i=1}^{n} \left(\tau_i s + 1\right)}
\tag{3.82}
$$

which gives the time constants, τ, of the process. If there are multiple time constants, then the largest time constant will determine the overall response speed of the process. Should the time constant be a complex number, then the speed of the response corresponds solely to its real component.

The final parameter of interest is the **time delay**, θ, which measures how long it takes before the system responds to a change in the input. Time delays can arise from two different sources: physical and measurement. Physically, time delays are perceived times at which a system is not responding to changes in the input. They can arise if it takes time for an observable change to be seen, for example, heating a large tank would take some time before the temperature rises by an appreciable amount that can be measured. On the other hand, measurement time delays arise due to the placement of sensors. In many systems, it may not be possible to measure the variable immediately where it will act on the process. In this case, it will take some time before the variable actually affects the system, for example, the flow rate in a pipe could be measured at the beginning of the pipe and it could take some time before it will reach the process. In the frequency domain, the time delay is found as

$$\mathrm{e}^{-\theta s} \tag{3.83}$$

where θ is the time delay. In many applications, it may be necessary to convert the exponential form of the time delay into a polynomial expansion. This conversion can be accomplished using the **Padé approximation**, denoted as the n/m Padé approximation, where n is the order of the polynomial in the numerator and m is the order of the polynomial in the denominator. The **1/1 Padé approximation** is given as

$$\mathrm{e}^{-\theta s} = \frac{1 - \frac{\theta}{2}s}{1 + \frac{\theta}{2}s} \tag{3.84}$$

The **2/2 Padé approximation** is given as

$$\mathrm{e}^{-\theta s} = \frac{1 - \frac{\theta}{2}s + \frac{\theta^2}{12}s^2}{1 + \frac{\theta}{2}s + \frac{\theta^2}{12}s^2} \tag{3.85}$$

A table of the Padé Approximation for the exponential function up to 3/3 are given in Table 3.5, from which the Padé Approximation for the time delay can be easily derived.

Table 3.5 Padé approximations for the exponential function, e^x, where $x = -\theta s$

$n\rightarrow$ $m\downarrow$	0	1	2	3
0	$\frac{1}{1}$	$\frac{1}{1-x}$	$\frac{1}{1-x+\frac{1}{2}x^2}$	$\frac{1}{1-x+\frac{1}{2}x^2-\frac{1}{6}x^3}$
1	$\frac{1+x}{1}$	$\frac{1+\frac{1}{2}x}{1-\frac{1}{2}x}$	$\frac{1+\frac{1}{3}x}{1-\frac{2}{3}x+\frac{1}{6}x^2}$	$\frac{1+\frac{1}{4}x}{1-\frac{3}{4}x+\frac{1}{4}x^2-\frac{1}{24}x^3}$
2	$\frac{1+x+\frac{1}{2}x^2}{1}$	$\frac{1+\frac{2}{3}x+\frac{1}{6}x^2}{1-\frac{1}{3}x}$	$\frac{1+\frac{1}{2}x+\frac{1}{12}x^2}{1-\frac{1}{2}x+\frac{1}{12}x^2}$	$\frac{1+\frac{2}{5}x+\frac{1}{20}x^2}{1-\frac{3}{5}x+\frac{3}{20}x^2-\frac{1}{60}x^3}$
3	$\frac{1+x+\frac{1}{2}x^2+\frac{1}{6}x^3}{1}$	$\frac{1+\frac{3}{4}x+\frac{1}{4}x^2+\frac{1}{24}x^3}{1-\frac{1}{4}x}$	$\frac{1+\frac{3}{5}x+\frac{3}{20}x^2+\frac{1}{60}x^3}{1-\frac{2}{5}x+\frac{1}{20}x^2}$	$\frac{1+\frac{1}{2}x+\frac{1}{10}x^2+\frac{1}{120}x^3}{1-\frac{1}{2}x+\frac{1}{10}x^2-\frac{1}{120}x^3}$

Example 3.9: Extracting Information from a Transfer Function
Determine the gain, time constant, and time delay for the following transfer function

$$G(s) = \frac{Y(s)}{U(s)} = \frac{1}{5s+3}e^{-5s} \tag{3.86}$$

Solution

Re-arranging into the required form gives

$$G(s) = \frac{Y(s)}{U(s)} = \frac{1/3}{\frac{5}{3}s+1}e^{-5s} \tag{3.87}$$

from which, by inspection, the gain is 1/3 and the time constant is 5/3. The time delay is 5.

3.3.1 Frequency-Domain Analysis

In frequency-domain analysis, the transfer function is solely used and analysed to determine its behaviour at different frequencies. Frequency-domain analysis is often presented graphically and was once used extensively before the advent of computers. Understanding the principles of frequency-domain analysis is still important, especially when designing various filters for electronic systems, such as microphones.

Frequency-domain methods are based on setting $s = j\omega$, where j is the imaginary number $\sqrt{1}$ and ω is the frequency. This basically means that we are considering the response of the system to sinusoidal waves with different frequencies. The goal is to see how the process changes the amplitude or strength and the phase shift. Assuming that the original input has the form

$$\sin \omega t \tag{3.88}$$

the response will have the form

$$A \sin(\omega t + \phi) \tag{3.89}$$

where A is the amplitude and ϕ is the phase shift. Consider a general transfer function, $G(s)$, then define the **amplitude ratio, AR**, (which corresponds to a normalised amplitude) as

$$\mathrm{AR} = \|G(j\omega)\| = \sqrt{\mathrm{Re}(G(j\omega))^2 + \mathrm{Im}(G(j\omega))^2} \tag{3.90}$$

where Re is the real component of a complex number, Im is the imaginary component, and $\|\cdot\|$ is the modulus function. It should be noted that in many applications, the logarithm of the amplitude ratio is used. In most cases, a base-10 logarithm is used. Occasionally, a decibel logarithm will be used, that is, $\mathrm{AR} = 20\log||G(j\omega)||$. The **phase angle**, ϕ, conventionally expressed in degrees, is defined as

$$\phi = \arctan\left(\frac{\mathrm{Im}(G(j\omega))}{\mathrm{Re}(G(j\omega))}\right) \tag{3.91}$$

where *arctan* is the standard inverse tangent function defined on]−90°, 90°[. It should be noted that if the denominator is negative, then it is necessary to add 180° to the value obtained previously. This is because the *arctan* function does not work on the complete circle.[8]

It can be noted that if the original transfer function is a product of simpler transfer functions, that is,

$$G(s) = \prod G_i(s) \tag{3.92}$$

then the amplitude ratio can be computed as

$$\mathrm{AR} = \prod \mathrm{AR}_{G_i(s)} \tag{3.93}$$

which is the product of the individual amplitude ratios corresponding to each simpler transfer function. The phase angle can be computed as

$$\phi = \sum \phi_{G_i(s)} \tag{3.94}$$

These two formulae can simplify the determination of the amplitude ratio and phase angles of complex transfer functions. This formulae result from the fact that all transfer functions can be expressed in polar form as $G(j\omega) = ||G(j\omega)||\mathrm{e}^{-\phi j}$, from which it is easy to see that combining multiple transfer functions as given by Eq. (3.92), the given formulae will result.

The amplitude ratio and phase angle can be graphically presented in two different forms, either as a **Bode plot** or as a **Nyquist plot**. In a Bode plot, the x-axis is the frequency, ω, and the y-axis is either the amplitude ratio or phase angle. A typical Bode plot is shown in Fig. 3.2.

In a Nyquist plot, the real component of $G(j\omega)$ is plotted on the x-axis and the imaginary component of $G(j\omega)$ is plotted on the y-axis. A typical Nyquist plot is shown in Fig. 3.3. In most applications, Bode plots are used, but a Nyquist plot can be useful to analyse the interactions between different systems.

[8] In some programmes, an "astronomical" arctangent function is defined as as *arctan2*(y, x), which will return the value in the correct quadrant based on the signs of y and x. If available, it should be used.

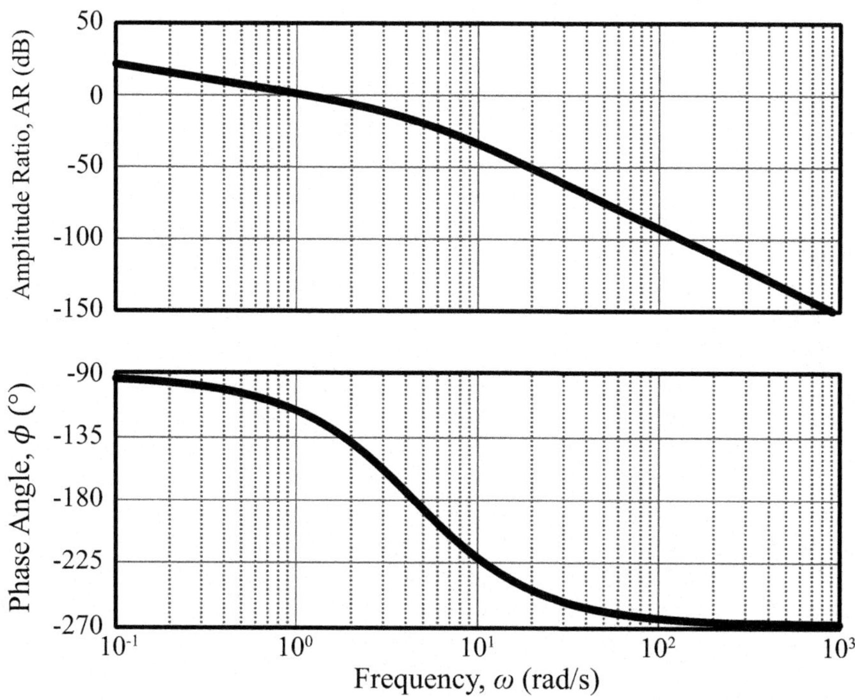

Fig. 3.2 Bode plot: The argument-ratio and phase-angle plots

Fig. 3.3 A Nyquist plot (for the same process as given by the Bode plot in Fig. 3.2)

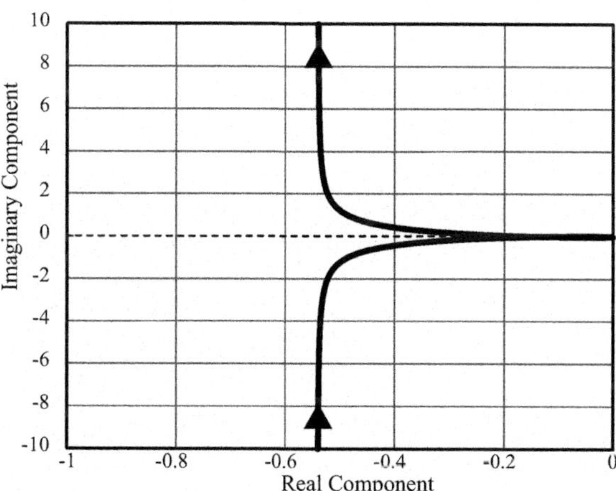

3.3.2 Stability

Stability of a model refers to the behaviour of the model as $t \rightarrow \infty$ for the case where the input is bounded, that is, $|u_t| \leq L$, where L is a positive constant. A model is said to be **stable** if as $t \rightarrow \infty$ for a bounded input, $y_\infty \rightarrow K$, where K is a constant. Otherwise, a model is said to be **unstable**. Determining the stability of a process depends on the type of representation used and nature of the time domain. Table 3.6 summarises the key results regarding stability. A detailed examination of how the different types of stability manifest themselves can be found in the Sect. Summary of the Functional Behaviour in Continuous- and Discrete-Time Domains.

For a state-space representation, stability is determined by examining the eigenvalues of the state matrix, $\mathcal{A}$. In the continuous-time domain, the real part of all the eigenvalues must be less than zero, for the model to be stable; otherwise, the model is said to be unstable. In the discrete-time domain, the magnitude of all the eigenvalues must be less than 1, that is, the eigenvalues must lie inside the unit circle, for the process to be stable; otherwise, the model is said to be unstable.

For a transfer-function representation, stability is determined by examining the poles, or roots of the denominator, D. In the continuous-time domain, the real part of all the poles must be less than zero, for the model to be stable; otherwise, the model is unstable. In the discrete-time domain, the magnitude of all the poles must be less than 1, that is, the poles must lie inside the unit circle, for the process to be stable; otherwise, the process is unstable.[9]

Finally, it can be noted that if the state-space representation is stable, then so will the transfer function representation be. However, the converse is not true, since each transfer function does not have a unique state-space representation and it is possible to construct an unstable state-space representation for a stable transfer function, by adding an unobservable unstable state.

Example 3.10: Determining the Stability of a Transfer Function
Determine if the following processes are stable:

$$G_1(s) = \frac{1}{5s+3}\mathrm{e}^{-5s} \tag{3.95}$$

$$G_2(s) = \frac{1}{(5s+3)\left(s^2+4\right)}\mathrm{e}^{-5s} \tag{3.96}$$

[9] In the discrete domain with a transfer function, it is very important to take into consideration what variable is being used and how the stability condition is being phrased. Often, especially in control, stability is discussed in terms of z^{-1}, in which case the above rules need to be inverted, that is, stability implies poles outside the unit circle. For consistency, stability in this book will be discussed in terms of z.

$$G_3(z) = \frac{z^2}{z^4 + z^3 + z^2 + z - 1} \tag{3.97}$$

Solution

For G_1, setting $5s+3$ equal to zero and solving, gives a pole of $-3/5$. According to the results in Table 3.6 for a continuous transfer function, the system is stable, since the pole is less than zero.

For G_2, setting the two terms to zero and solving, gives three poles of $-3/5$ and $\pm 2i$. From Table 3.6 for a continuous transfer function, we conclude that the system is unstable since we have two poles with a real component equal to zero and one pole whose value is less than zero.

For G_3, the setting the denominator equal to zero and solving, gives poles of $-1.291, -0.1141 \pm 1.217i$, and 0.519. For a discrete transfer function, Table 3.6 states that in order for the system to be stable, the magnitude of the roots must be less than one. We see that we have at least one pole (-1.291) whose magnitude is greater than zero. This implies that the system is unstable.

Example 3.11: Determining the Stability of a State-Space Model
Determine if the given $\mathcal{A}$-matrices represent stable continuous processes:

$$\mathcal{A}_1 = \begin{bmatrix} 2 & 0 \\ 3 & -2 \end{bmatrix} \tag{3.98}$$

$$\mathcal{A}_2 = \begin{bmatrix} -5 & 5 & 6 \\ 0 & -3 & 2 \\ 0 & 0 & -1 \end{bmatrix} \tag{3.99}$$

$$\mathcal{A}_3 = \begin{bmatrix} 2 & 1 \\ -1 & 2 \end{bmatrix} \tag{3.100}$$

Solution

For $\mathcal{A}_1$, since it is a triangular matrix, the eigenvalues can be directly found by looking at the main diagonal entries. This implies that the eigenvalues are 2 and -2. According to the results in Table 3.6 for a continuous state-space model, the system is unstable, since one of the eigenvalues is greater than zero.

Likewise, for $\mathcal{A}_2$, we can find the eigenvalues by looking at the main diagonal entries. This gives $-5, -3$, and -1. Since all of the eigenvalues are less than zero, from Table 3.6 for a continuous state-space model, we conclude that the system is stable.

For $\mathcal{A}_3$, we need to compute the eigenvalues as follows

$$\begin{aligned}\det(\lambda I - \mathcal{A}_3) &= (\lambda - 2)^2 - (1)(-1) \\ &= \lambda^2 - 4\lambda + 5 \\ &= 2 \pm i\end{aligned} \tag{3.101}$$

Note that the real part of the eigenvalues is greater than zero. From Table 3.6, we see that the system is stable if all real parts of the poles are less than zero. Since this is not the case here, we conclude that the system is unstable.

Routh Stability Analysis

Although the results presented in Table 3.6 are very useful when a numerical transfer function is available, determining the roots algebraically can be difficult, if not impossible for higher-order systems. For this reason, there exist special tests for stability that do not require the roots of the polynomial to be determined. One of the most common tests for a continuous-time transfer function is the **Routh stability analysis**. By simply considering the coefficients of the polynomial of the denominator, the stability of the system can be determined. Consider the characteristic polynomial (the polynomial of the denominator) as

$$a_n s^n + a_{n-1} s^{n-1} + a_{n-2} s^{n-2} + \ldots + a_1 s + a_0 = 0 \tag{3.102}$$

There are two conditions for the Routh stability test:

(1) All coefficients must be present and have the same sign (either all strictly positive, that is, greater than zero or all strictly negative, that is less than zero). If this is not the case, the system is unstable.
(2) If Condition 1 is satisfied, then create the table shown in Table 3.7.

Table 3.6 Summary of the stability conditions for different representations and time domains

Representation		Analysis metric, p	Stable	Unstable
Continuous time	State-space	Eigenvalues of the state-space matrix, $\mathcal{A}$	$\mathrm{Re}(p_i)<0, \forall i$	$\mathrm{Re}(p_i)\geq 0$
	Transfer function	Poles of the transfer function (Roots of D)	$\mathrm{Re}(p_i)<0, \forall i$	$\mathrm{Re}(p_i)\geq 0$
Discrete time	State-space	Eigenvalues of the state-space matrix, $\mathcal{A}$	$\lVert p_i \rVert < 1, \forall i$	$\lVert p_i \rVert \geq 1$
	Transfer function	Poles of the transfer function (Roots of D)	$\lVert p_i \rVert < 1, \forall i$	$\lVert p_i \rVert \geq 1$

Table 3.7 Analysis table for Routh stability

Row	1	2	3	…
1	a_n	a_{n-2}	a_{n-4}	…
2	a_{n-1}	a_{n-3}	a_{n-5}	…
3	$b_1 = \frac{a_{n-1}a_{n-2} - a_n a_{n-3}}{a_{n-1}}$	$b_2 = \frac{a_{n-1}a_{n-4} - a_n a_{n-5}}{a_{n-1}}$	…	
4	$c_1 = \frac{b_1 a_{n-3} - b_2 a_{n-1}}{b_1}$	…		
⋮	⋮	⋮		
$n+1$	z_1			

(a) In the first row, place the coefficients corresponding to a_n, a_{n-2},…

(b) In the second row, place the coefficients corresponding to a_{n-1}, a_{n-3},…

(c) Using the values from the rows above and the current column and the column to the right, compute a determinant-like number using the formula provided in the table. For the ith row in the jth column ($i \in \{1, 2, \ldots, n+1$, $j \in \{1, 2, \ldots, \lceil 0.5n \rceil\}$), the general formula is

$$\lambda_{i,j} = \frac{\lambda_{i-2,j+1}\lambda_{i-1,j} - \lambda_{i-2,j}\lambda_{i-1,j+1}}{\lambda_{i-1,1}} \tag{3.103}$$

where $\lambda_{1,j} = a_{n-2j+2}$ and $\lambda_{2,j} = a_{n-2j+1}$ (the first two rows are the coefficients of the characteristic polynomial). Any values that are not given are assumed to be zero.

(d) Continue computing this value until the table has $n+1$ rows.

(e) A polynomial is stable (all roots of the characteristic polynomial have negative real components) if all the values in the first column have the same sign, either strictly positive or strictly negative.

Example 3.12: Example of Routh Stability Analysis

Using Routh stability, determine the stability of the following two transfer functions:

(1) $G(s) = \dfrac{5}{s^6 + 5s^4 - 6s^3 + 2s^2 + 3s + 1}$

(2) $G(s) = \dfrac{4}{s^4 + 3s^3 + s^2 + 2s + 1}$

Solution

For the first transfer function, we can determine the stability by inspection. Since we are missing one of the powers (s^5) and the coefficients are both negative and positive, we can conclude on the basis of Condition (1) of the Routh stability analysis that this transfer function is unstable.

For the second transfer function, we see that Condition (1) is satisfied, as all the powers are present and the coefficients have the same sign. Therefore, we will need to check condition (2) by constructing the Routh array. There will be 3 (always $n / 2$, rounded up) columns in the array. In the first row (denoted as **1** in Table 3.8), we place the even coefficients ordered in decreasing powers. In the second row, we place the odd coefficients. In Rows 3, 4, and 5, we compute the values using the formula:

$$b_1 = \frac{a_{n-1}a_{n-2} - a_n a_{n-3}}{a_{n-1}} = \frac{3(1) - 1(2)}{3} = \frac{1}{3}$$

$$b_2 = \frac{a_{n-1}a_{n-4} - a_n a_{n-5}}{a_{n-1}} = \frac{3(1) - 1(0)}{3} = 1$$

$$c_1 = \frac{b_1 a_{n-3} - b_2 a_{n-1}}{b_1} = \frac{(1/3)2 - 3(1)}{1/3} = -7$$

$$z_1 = \lambda_{5,1} = \frac{\lambda_{3,2}\lambda_{4,1} - \lambda_{3,1}\lambda_{4,2}}{\lambda_{4,1}} \frac{-7(1) - (1/3)(0)}{-7} = 1$$

It can be noted that any values that do not exist (such as a_{n-5}) are equal to zero. As stated by the definition of the condition, we will always have $n+1$ rows. The formulae used to compute the subsequent values are essentially determinants with the sign in the numerator flipped.

From Table 3.8, we see that there is a sign change in Column 1, Row 4, which implies that the system is unstable.

Table 3.8 Routh array

Row	1	2	3
1	1	1	1
2	3	2	
3	1/3	1	
4	−7		
5	1		

Jury Stability Analysis

Although the results presented in Table 3.6 are very useful when a numerical transfer function is available, determining the roots algebraically can be difficult, if not impossible for higher-order systems. For this reason, there exist special tests for stability that do not require the roots of the polynomial to be determined. One

Table 3.9 Analysis table for Jury stability

Row	0	1	…	$n-1$	n
1	a_0	a_1	…	a_{n-1}	a_n
2	a_n	a_{n-1}	…	a_1	a_0
3	$b_0 = a_0 a_0 - a_n a_n$	$b_1 = a_0 a_1 - a_n a_{n-1}$	…	$b_{n+1} = a_0 a_{n-1} - a_n a_1$	0
4	b_{n-1}	b_{n-2}	…	b_0	0
5					
⋮	⋮	⋮	⋮	⋮	⋮
$2n-3$	$\boldsymbol{q_0}$	$\boldsymbol{q_1}$	**…**	**…**	**0**

of the most common tests for a discrete-time transfer function is the **Jury stability analysis**. By simply considering the coefficients of the polynomial of the denominator, the stability of the system can be determined. Like with the Routh stability test, a table is constructed. Let the polynomial of interest be given as

$$D(z) = a_n z^n + a_{n-1} z^{n-1} + \cdots + a_1 z + a_0 \tag{3.104}$$

Construct the following table (similar to that shown in Table 3.9), where the first row contains all the coefficients starting from a_0 and ending with a_n. The second row is the first row reversed, that is, the first column contains a_n, and the last column contains a_0. The third row is computed using

$$b_i = a_0 a_i - a_n a_{n-i} \tag{3.105}$$

while the fourth row is the third row reversed. Note that only the first n values are reversed. Effectively, we have now created a new reduced polynomial that needs to be analysed. Thus, the fifth row is computed using the same formula as for the third row given by Eq. (3.105), but replacing n by $n-1$ coefficients, that is,

$$c_i = b_0 b_i - b_{n-1} b_{n-1-i} \tag{3.106}$$

This procedure is then repeated until there is a row with only 3 elements left (denoted as q_0, q_1, and q_2), that is, the table will have $2n-3$ rows. In general, for the $(2j+1)$-th row ($j = 1, 2, \ldots, n-2$), that is, each odd-numbered row, we can write the relationship as follows

$$d_i^{(2j+1)} = d_0^{(2j-1)} d_i^{(2j-1)} - d_{n-j+1}^{(2j-1)} d_{n-j-i+1}^{(2j-1)} \tag{3.107}$$

where $d_i^{(1)} = a_i$, that is, the first row contains the actual polynomial coefficients. The $(2j+2)$-th row, that is, the even row, will then be the $n-j+1$ coefficients written in reversed order. It can be noted that the odd rows are essentially a determinant of the above two rows.

The conditions for stability can then be stated as:

(1) $D(1) > 0$
(2) $(-1)^n D(-1) > 0$
(3) $|a_0| > |a_n|$
(4) $|b_0| > |b_{n-1}|$, and continuing in this manner until the last row is reached, where $|q_0| > |q_2|$.

If any of the above conditions fail, then the system will have at least one pole outside the unit circle and hence be unstable.

Example 3.13: Example of Jury Stability Analysis
Using the Jury stability test, determine the stability of the following two discrete transfer functions:

(1) $G(z) = \frac{5}{z^6+5z^4+6z^3+2z^2+3z+1}$
(2) $G(z) = \frac{4}{10z^4+3z^3+z^2+5z+1}$

Solution

For the first transfer function, we need to test the initial constraints before we consider creating the table. For Condition 1, $D(1) > 0$, we get

$$1+5+6+2+3+1 = 18 > 1$$

which implies that this condition is satisfied. Condition 2, $(-1)^n D(-1) > 0$, gives

$$(-1)^6\left[1(-1)^6+5(-1)^4+6(-1)^3+2(-1)^2+3(-1)^1+1\right] = 0$$

which is not satisfied. This implies that the first transfer function is not stable.

For the second transfer function, testing the three initial conditions gives

Condition 1: $10+3+1+5+1 = 20 > 0$ (satisfied)

Condition 2: $(-1)^4\begin{bmatrix}10(-1)^4+3(-1)^3\\+(-1)^2+5(-1)^1+1\end{bmatrix} = 4 > 0$ (satisfied)

Condition 3: $|a_n| > |a_0| \Rightarrow |10| > |1|$ (satisfied)

Since all the preliminary conditions have been satisfied, the Jury array can be constructed. This is shown in Table 3.10. The first row consists of the coefficients arranged in increasing powers, while the second row consists of the coefficients arranged in decreasing power. In Row 3, we compute the values using the formula

Table 3.10 Table for Jury stability

Row	0	1	2	3	4
1	1	5	1	3	10
2	10	3	1	5	1
3	−99	−25	−9	−47	
4	−47	−9	−25	−99	
5	7592	2052	−284		

$$b_i = a_0 a_i - a_n a_{n-i} \tag{3.108}$$

which gives

$$b_0 = a_0 a_0 - a_4 a_{4-0} = 1(1) - 10(10) = -99$$

$$b_1 = a_0 a_1 - a_4 a_{4-1} = (1)5 - 10(3) = -25$$

$$b_2 = a_0 a_2 - a_4 a_{4-2} = 1(1) - 10(1) = -9$$

$$b_3 = a_0 a_3 - a_4 a_{4-3} = 3(1) - 10(5) = -47$$

From here, we should check that the condition $|b_0| > |b_{n-1}|$ is satisfied. Since $|-99| > |-47|$, it holds and we proceed to the next step. We reverse the order of the remaining $n - 1$ coefficients and write them in Row 4. Effectively, we have created a new polynomial of order $n - 1$ that we need to test. Thus, in Row 5, we will use a modified form of Eq. (3.108) to give

$$c_i = b_0\, b_i - b_{n-1} b_{n-1-i} \tag{3.109}$$

Effectively, this is the same as Eq. (3.108), but with the value of n reduced by 1. Evaluating Eq. (3.109) for the $n - 2$ columns gives

$$c_0 = b_0 b_0 - b_3 b_3 = (-99)^2 - (-47)^2 = 7592$$

$$c_1 = b_0 b_1 - b_3 b_2 = -99(-25) - (-47)(-9) = 2052$$

$$c_2 = b_0 b_2 - b_3 b_1 = -9(-99) - (-47)(-25) = -284$$

Since we are left with three columns, we simply need to determine $|q_0| > |q_2|$, which in our case becomes $|7592| > |-284|$. Since it holds, we can conclude that the system is stable.

Closed-Loop Stability Analysis

For closed-loop systems, stability can be analysed by considering the closed-loop transfer function, G_{cl},

$$G_{cl} = \frac{G_c G_p}{1 + G_c G_p} \tag{3.110}$$

Since it has been shown that the poles of the system are the determining factor for stability, it therefore suffices to consider only the denominator of the closed-loop transfer function, that is, the term $1 + G_c G_p$. It is possible to apply either the Routh or the Jury stability tests to determine under what conditions the system will be stable.

However, in some circumstances, it can be instructive to look at the frequency-domain plots, especially the Bode plot, to determine stability. In order to make the analysis simpler, we will rewrite the denominator to

$$G_c G_p = -1 \tag{3.111}$$

Taking the magnitude (or modulus) of Eq. (3.111) gives

$$\left\|G_c G_p\right\| = \|-1\| \tag{3.112}$$

while the phase angle for -1 is $-180°$, since the imaginary component is 0 and the real component is -1, which after subtracting $-180°$ gives the correct location. Therefore, plotting the Bode plot for the open-loop transfer function $G_c G_p$ will allow the stability of the system to be determined by examining the **critical frequency**, ω_c, which is the frequency corresponding to a phase angle of $-180°$. If the value of the amplitude ratio at this point is greater than 1 (or 0 if a logarithmic basis is being used), then the process will be closed-loop unstable. An example is shown in Fig. 3.4.

Furthermore, we can define two additional parameters of interest: the **gain margin** (GM) and the **phase margin** (PM). These two margins represent how much room we have before the system becomes unstable. On a Bode plot, instability implies an amplitude ratio above 1 (or 0 for the logarithmic case) or a phase angle below $-180°$. These parameters are shown in Fig. 3.5 for the Bode plot and in Fig. 3.6 for the Nyquist plot. The phase margin is defined by adding 180° to the phase angle when the gain crosses the value of 1, which is denoted by ω_g and called the gain crossover, that is,

$$\text{PM} = \phi\left(\omega_g\right) + 180° \tag{3.113}$$

The gain margin is defined as the inverse of the amplitude ratio when the phase angle crosses the phase angle of $-180°$, which is denoted by ω_p and called the phase crossover, that is,

$$\text{GM} = 1/\text{AR}\left(\omega_p\right) \tag{3.114}$$

It is possible to design controllers by specifying the phase and gain margins.

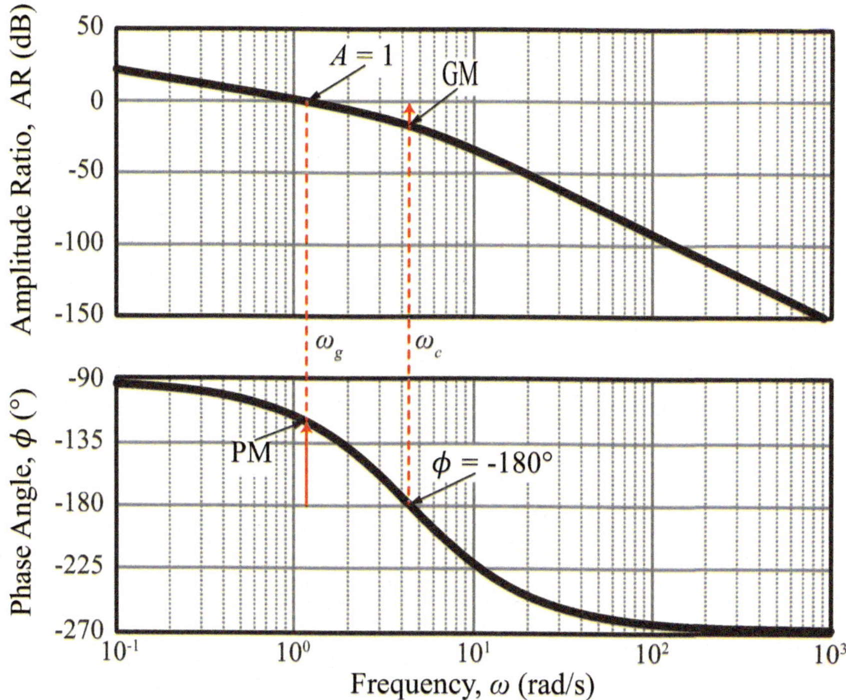

Fig. 3.4 Bode plot for closed-loop stability analysis

3.3.3 *Realisability*

The **realisability** of a transfer considers whether a given transfer function can be implemented in a real system. There are three conditions that need to be fulfilled for a transfer function to be realisable:

(1) **Stability**: The transfer function should be stable. If there be any roots on the imaginary axis for the continuous case or on the unit zero for the discrete case, then they should not have a multiplicity greater than one. This implies that the integrator and the sine function can be implemented.
(2) **Causality**: The transfer function should be causal.
(3) **Finite Frequency Response**: The frequency response of the function should be finite, that is,

$$0 \leq \lim_{\omega \to \infty} \mathrm{AR}(\omega) = K < \infty \tag{3.115}$$

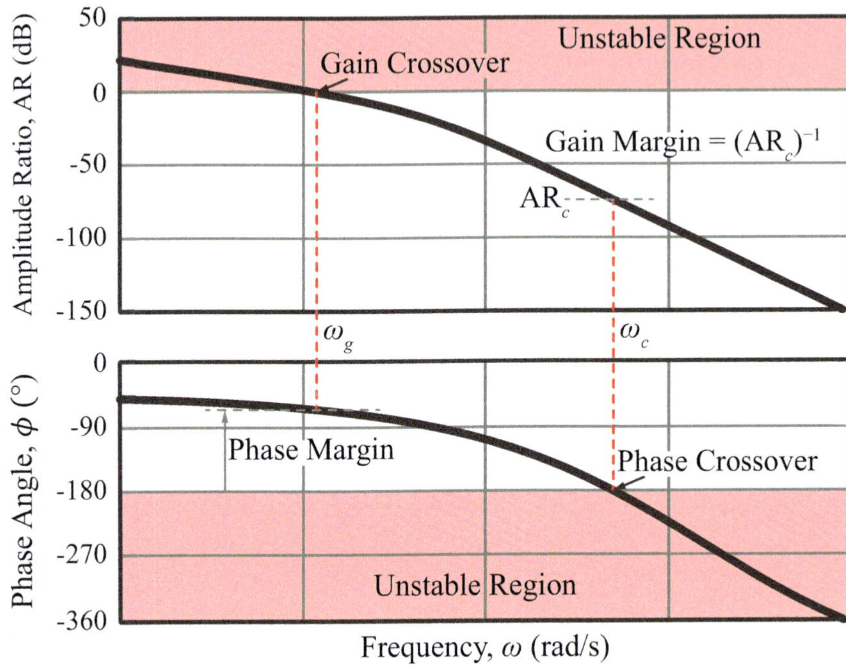

Fig. 3.5 Closed-loop stability using the Bode plot

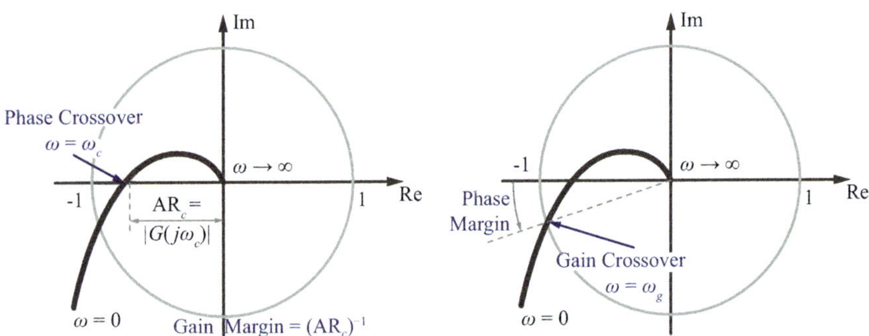

Fig. 3.6 Closed-loop stability using the Nyquist plot

Since the rules for determining stability and causality are known, only the last condition needs to be determined. For a continuous-time transfer function, the frequency response will be finite if the transfer function is proper. For a discrete-time transfer function, this condition is not relevant, since all rational transfer function will have a finite frequency response.

3.3.4 Controllability and Observability

Since we are ultimately interested in automating our process, it is important to understand if our process can in fact be automated. For state-space representations, it is very common to talk about the controllability and observability of the process. The state equation or the pair $(\mathcal{A}, \mathcal{C})$ is **observable** if and only if for any unknown initial state, x_0, there exists a finite time t (greater than zero), such that the knowledge of the input, $u(t)$, and output, $y(t)$ in the interval $[0, t]$ is sufficient to determine the initial state, x_0. This can be accomplished if and only if

(1) $$\mathcal{O} = \begin{bmatrix} \mathcal{C} \\ \mathcal{C}\mathcal{A} \\ \mathcal{C}\mathcal{A}^2 \\ \vdots \\ \mathcal{C}\mathcal{A}^{n-1} \end{bmatrix}_{nq\times n}$$ has full rank n, where $\mathcal{O}$ is the observability matrix.

(2) The observability Grammian, $\mathcal{W}_o(t)_{n\times n} = \int_0^t e^{\mathcal{A}^T\tau}\mathcal{C}^T\mathcal{C}e^{\mathcal{A}\tau}d\tau$ is nonsingular for all $t > 0$. The energy of the output signal can be given as $E_y(t) = x^T(0)\mathcal{W}_c(t)x(0)$.

(3) The $n\times n$ matrix $\left[\dfrac{\mathcal{A} - \lambda_i\mathcal{I}}{\mathcal{C}}\right]$ has rank n for all eigenvalues λ_i of $\mathcal{A}$. This allows the individual states or eigenvalues of $\mathcal{A}$ to be classified.

A system is said to be **detectable** if all unstable states of a state-space representation can be observed. This can be determined by examining $\left[\dfrac{\mathcal{A} - \lambda_i\mathcal{I}}{\mathcal{C}}\right]$ for each of the unstable states and determining if it is full rank.

The state equation or the pair $(\mathcal{A}, \mathcal{B})$ is said to be **controllable** if and only if for any initial state, x_0, and final state, x_1, there exists a bounded input, $u(t)$, that transfers x_0 to x_1 in finite time. This can be accomplished if and only if:

(1) $\overline{\mathcal{C}}_{n\times np} = \begin{bmatrix} \mathcal{B} & \mathcal{A}\mathcal{B} & \mathcal{A}^2\mathcal{B} & \cdots & \mathcal{A}^{n-1}\mathcal{B} \end{bmatrix}$ has full rank n, where $\overline{\mathcal{C}}$ is the controllability matrix.

(2) The controllability Grammian, $\mathcal{W}_c(t)_{n\times n} = \int_0^t e^{\mathcal{A}\tau}\mathcal{B}\mathcal{B}^T e^{\mathcal{A}^T\tau}d\tau$, is nonsingular for all $t > 0$. $\mathcal{W}_c^{-1}$ is a measure of the energy required to control the process. A larger value implies that more energy (or effort) is required.

(3) The $n\times n$ matrix $\left[\mathcal{A} - \lambda_i\,\mathcal{J} \middle| \mathcal{B}\right]$ has rank n for all eigenvalues λ_i of $\mathcal{A}$. This allows the individual states or eigenvalues of $\mathcal{A}$ to be classified.

A system is said to be **stabilisable** if all unstable states of a state-space representation can be controlled. This can be determined by examining $\left[\mathcal{A} - \lambda_i \mathcal{I} | \mathcal{B}\right]$ for each of the unstable states and determining if it is full rank.

It can be noted that controllability and observability are duals of each other, that is, if $(\mathcal{A}, \mathcal{B})$ is controllable, then $(\mathcal{A}^T, \mathcal{B}^T)$ is observable. Similarly, if $(\mathcal{A}, \mathcal{C})$ is observable, then $(\mathcal{A}^T, \mathcal{C}^T)$ is controllable. This also explains why the controllable and observable canonical forms are transposes of each other.

3.3.5 Analysis of Special Transfer Functions

In this section, the properties of different commonly encountered transfer functions will be considered. Since much of the analysis in automation focuses on using transfer functions, it is important to understand the behaviour of the common transfer functions. Time-domain responses of these transfer functions to a step input will also be considered, since it is important to recognise these transfer functions in their most common manifestations in a real process. Discrete-time systems will only be briefly considered, since most discrete-time analysis is still based on the underlying continuous-time systems.

Integrator

The integrator, as its name suggests, integrates a variable. It is a common model for level in a tank. Its Laplace transform is

$$G_I = \frac{M_I}{s} \tag{3.116}$$

where M_I is the gain. It is an unstable system, which will always increase even if the input is bounded. The step response of the system can be determined as

$$\begin{aligned} Y &= G_I U \\ Y &= \frac{M_I}{s}\frac{M}{s} \\ &= \frac{M_I M}{s^2} \end{aligned} \tag{3.117}$$

From Table 3.1, the time domain representation for Eq. (3.117) is

$$y_t = M_I M t \tag{3.118}$$

A representative plot is shown in Fig. 3.7. Its Bode plot can be determined as

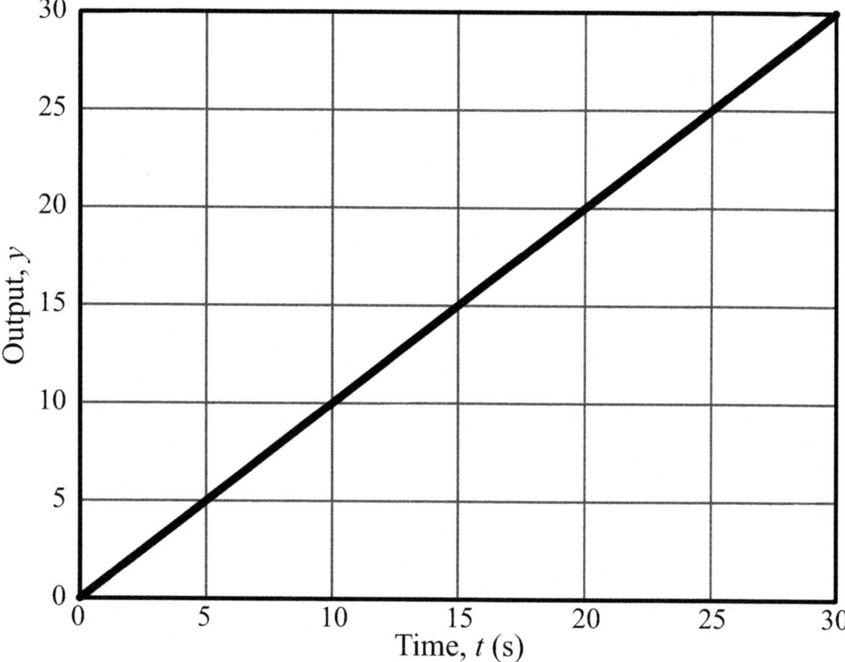

Fig. 3.7 Response of the integrator 1 / s to a unit-step input

$$
\begin{aligned}
G_I(j\omega) &= \frac{M_I}{j\omega} = -\frac{M_I j}{\omega} \\
\Rightarrow \text{AR} &= \sqrt{0^2 + \left(-\frac{M_I}{\omega}\right)^2} = \frac{|M_I|}{\omega} \\
\phi &= \arctan\left(\frac{-\frac{M_I}{\omega}}{0}\right) = -90^\circ
\end{aligned}
\tag{3.119}
$$

A representative Bode plot is shown in Fig. 3.8. The key properties of the integrator are summarised in Table 3.11.

Lead Term

The lead term is a simple, first-order transfer function that is occasionally found in combination with other transfer function to model unusual or more complex initial dynamics in a process. Its Laplace transform is

$$
G_L = K(\tau_L s + 1) \tag{3.120}
$$

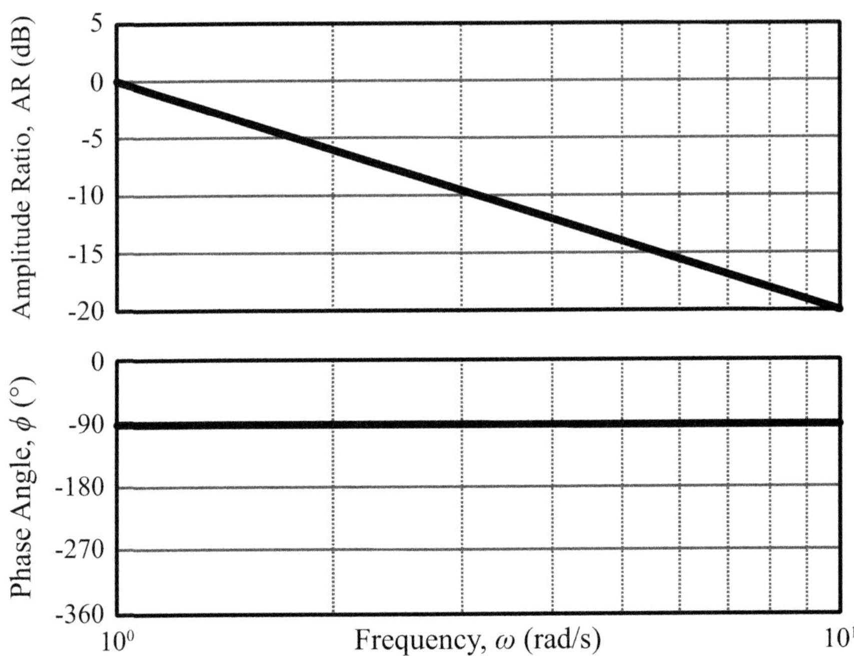

Fig. 3.8 Bode plot for an integrator

Table 3.11 Basic information about an integrator

Property		Value
Laplace transform		$G_I = \frac{M_I}{s}$
Step response (time domain)		$y_t = M_I M t$
Bode plots	AR	$\lvert M_I \rvert / \omega$
	ϕ	$-90°$
Stable		No

where K is the gain and τ_L is a time constant. The step response of the system can be determined as

$$\begin{aligned} Y &= G_L U \\ Y &= K(\tau_L s + 1)\frac{M}{s} \\ &= KM\frac{(\tau_L s + 1)}{s} \end{aligned} \tag{3.121}$$

In order to obtain a time-domain representation, it is necessary to perform a partial fraction decomposition of Eq. (3.121) to give

$$Y = KM\frac{(\tau_L s + 1)}{s}$$
$$Y = KM\left(\tau_L + \frac{1}{s}\right) \tag{3.122}$$

From Table 3.2, the time domain representation for Eq. (3.122) is

$$y_t = KM(\tau_L \delta + u_t) \tag{3.123}$$

where δ is the Dirac delta function and u_t is the unit step function. Its Bode plot can be determined as

$$\begin{aligned} G_L(j\omega) &= K(\tau_L j\omega + 1) = K + K\tau_L \omega j \\ \Rightarrow \text{AR} &= \sqrt{K^2 + (K\tau_L\omega)^2} = |K|\sqrt{1 + \tau_L^2\omega^2} \\ \phi &= \arctan\left(\frac{K\tau_L\omega}{K}\right) = \begin{cases} \arctan \tau_L\omega & K > 0 \\ \arctan \tau_L\omega + 180^\circ & K < 0 \end{cases} \end{aligned} \tag{3.124}$$

A representative Bode plot is shown in Fig. 3.9 for all 4 possible combinations of K and τ_L. The key properties of the lead term are summarised in Table 3.12.

First-Order System

A first-order system is one of the most common transfer functions encountered in automation engineering. It can be used to model any system ranging from simple heated tanks to complex multicomponent reactions. Its Laplace transform is

$$G_F = \frac{K}{(\tau_p s + 1)} \tag{3.125}$$

where K is the gain and τ_p is the process time constant. It is often coupled with a time-delay term to give a **first-order plus deadtime (FOPDT)** process model, that is,

$$G_{FD} = \frac{K}{(\tau_p s + 1)} e^{-\theta s} \tag{3.126}$$

The step response of the pure, first-order system can be determined as

$$\begin{aligned} Y &= G_F U \\ Y &= \frac{K}{(\tau_p s + 1)} \frac{M}{s} \\ &= \frac{KM}{s(\tau_p s + 1)} \end{aligned} \tag{3.127}$$

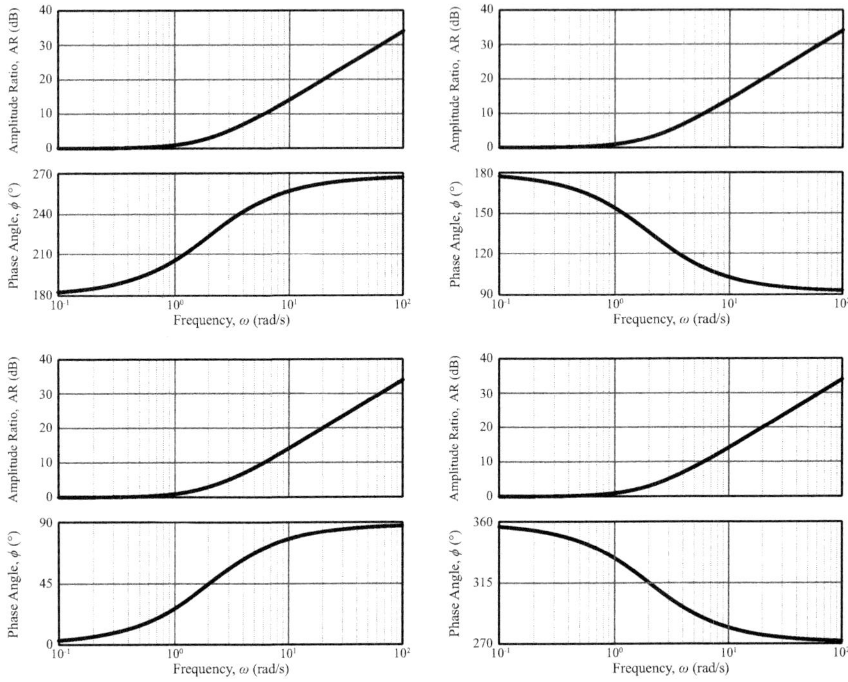

Fig. 3.9 Bode plot for a lead term (top row) $K > 0$, (bottom row) $K < 0$, (left) $\tau_L > 0$, and (right) $\tau_L < 0$

Table 3.12 Basic information about a lead term

Property		Value
Laplace transform		$G_L = K(\tau_L s + 1)$
Step response (time domain)		$y_t = KM(\tau_L \delta + u_t)$
Bode plots	AR	$\lvert K\rvert\sqrt{1+\tau_L^2\omega^2}$
	ϕ	$\begin{cases} \arctan \tau_L \omega & K > 0 \\ \arctan \tau_L \omega + 180^\circ & K < 0 \end{cases}$
Stable		Yes
Comment		Negative values of τ_L (positive zeros) can induce an inverse response in a system, that is, the variable first decreases in value and increases to reach its new steady-state value (or *vice versa*)

In order to obtain a time-domain representation, it is necessary to perform a partial-fraction decomposition of Eq. (3.127) to give

$$\begin{aligned} Y &= \frac{KM}{s(\tau_p s + 1)} \\ Y(s) &= KM\left(\frac{1}{s} + \frac{-\tau_p}{\tau_p s + 1}\right) \end{aligned} \tag{3.128}$$

From Table 3.2, the time domain representation for Eq. (3.128) is

$$y(t) = KM\left(1 - \mathrm{e}^{-\frac{t}{\tau_p}}\right) \tag{3.129}$$

The process is stable if $\tau_p > 0$, but unstable if $\tau_p < 0$. Assuming $\tau_p > 0$, then using the final-value theorem, it can be shown that the new steady-state value will be

$$\begin{aligned} \lim_{t\to\infty} y_t &= \lim_{s\to 0} sY(s) = \lim_{s\to 0} s\frac{MG_F(s)}{s} \\ &= \lim_{s\to 0} MG_F(s) = \lim_{s\to 0} MG_F(0) \\ &= KM \end{aligned} \tag{3.130}$$

The time response of a stable first-order system is shown in Fig. 3.10, as well as how to compute the key parameters from its graph. Its Bode plot can be determined as

$$\begin{aligned} G_F(j\omega) &= \frac{K}{(\tau_p j\omega + 1)} = \frac{K(1 - \tau_p \omega j)}{1 + \tau_p^2 \omega^2} \\ \Rightarrow \mathrm{AR} &= \sqrt{\frac{K^2 + (-K\tau_p\omega)^2}{(1 + \tau_p^2\omega^2)^2}} = \frac{|K|}{\sqrt{1 + \tau_p^2\omega^2}} \\ \phi &= \arctan\left(\frac{-K\tau_p\omega}{K}\right) = \begin{cases} -\arctan \tau_p\omega & K > 0 \\ 180° - \arctan \tau_p\omega & K < 0 \end{cases} \end{aligned} \tag{3.131}$$

In order to avoid a discontinuity when it passes through 0°, it is common to use negative angles when plotting this function, that is −45° rather than the equivalent 315°. A representative Bode plot is shown in Fig. 3.11 for all 4 possible combinations of K and τ_p. The properties of the first-order system are summarised in Table 3.13.

Second-Order System

A second-order system is another commonly encountered transfer function that is often used to model oscillations or periodic behaviour in a process. Its Laplace transform is

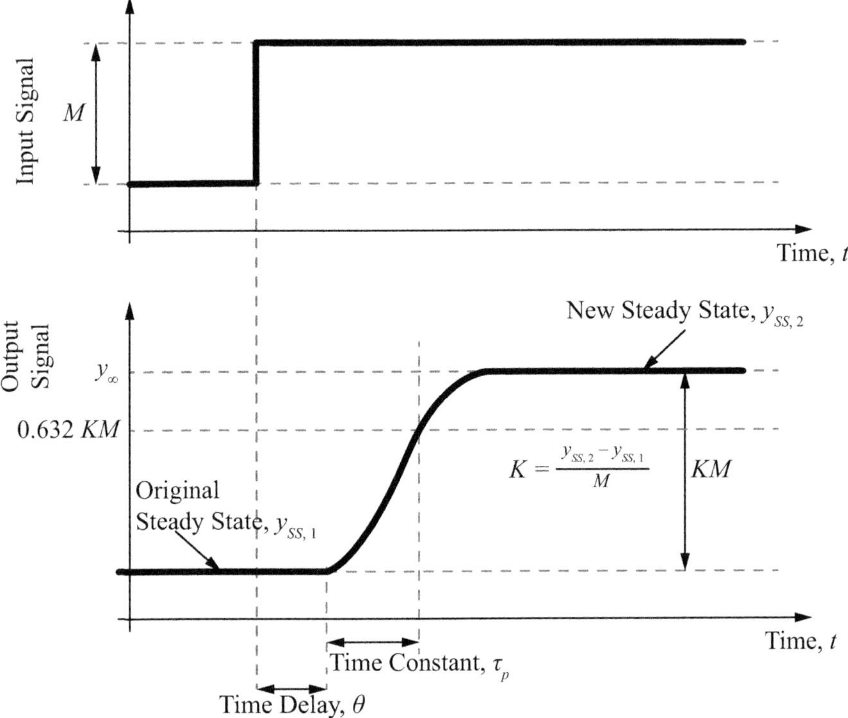

Fig. 3.10 Step response of a stable, first-order system

$$G_{II} = \frac{K}{\left(\tau_p^2 s^2 + 2\zeta\tau_p s + 1\right)} \tag{3.132}$$

where K is the gain, τ_p is the process time constant, and ζ is the damping coefficient. This basic transfer function can be augmented by adding a lead term to model various behaviours to give

$$G_{II} = \frac{K(\tau_L s + 1)}{\left(\tau_p^2 s^2 + 2\zeta\tau_p s + 1\right)} \tag{3.133}$$

As well, it can be coupled with a time-delay term to give a **second-order plus deadtime (SOPDT)** process model, that is,

$$G_{IID} = \frac{K}{\left(\tau_p^2 s^2 + 2\zeta\tau_p s + 1\right)} e^{-\theta s} \tag{3.134}$$

Table 3.13 Basic information about a first-order system

Property		Value
Laplace transform		$G_F = \frac{K}{(\tau_p s+1)}$
Step response (time domain)		$y(t) = KM\left(1 - \mathrm{e}^{-\frac{t}{\tau}}\right)$
Bode plots	AR	$\frac{\lvert K\rvert}{\sqrt{1+\tau_p^2\omega^2}}$
	ϕ	$\begin{cases} -\arctan \tau_p\omega & K > 0 \\ 180^\circ - \arctan \tau_p\omega & K < 0 \end{cases}$
Stable		Yes, if $\tau_p > 0$
Comment		This is one of the most common transfer functions encountered in automation engineering. It is used to model many different applications, often by including a time delay

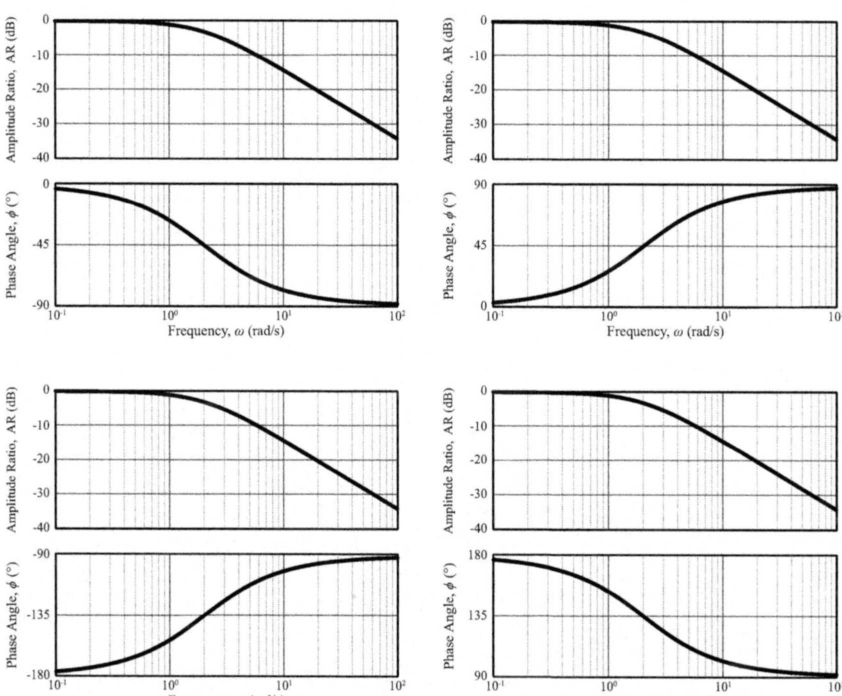

Fig. 3.11 Bode plot for a first-order system (top row) $K > 0$, (bottom row) $K < 0$, (left) $\tau_p > 0$, and (right) $\tau_p < 0$

The poles of the second-order transfer function can be written as

$$\begin{aligned} s &= \frac{-2\zeta\tau_p \pm \sqrt{4\zeta^2\tau_p^2 - 4\tau_p^2}}{2\tau_p^2} \\ &= \frac{-\zeta \pm \sqrt{\zeta^2 - 1}}{\tau_p} \end{aligned} \tag{3.135}$$

Depending on the value of ζ, three different cases can be determined:

(1) **Case I: Underdamped System**, where $|\zeta| < 1$. In this case, the poles will contain an imaginary component.
(2) **Case II: Critically Damped System**, where $|\zeta| = 1$. In this case, the poles will both be real and the same.
(3) **Case III:Overdamped System**, where $|\zeta| > 1$. In this case, the poles will both be real.

The behaviour and critical information depend on the particular case being considered.

For the **underdamped** case, where $|\zeta| < 1$, the poles of Eq. (3.132) will contain an imaginary component. The presence of an imaginary component will imply that there will be oscillations present in the time response. If $\zeta = 0$, then the system will continuously oscillate about a mean value. In all other cases, stability will depend on the sign of $\zeta\tau_p$: if positive, the system will be stable (decaying oscillations); if negative, the system will be unstable (increasing oscillations). The step response of this system can be written as

$$y(t) = KM\left(1 - \mathrm{e}^{-\zeta t/\tau_p}\left[\cos\left(\frac{\sqrt{1-\zeta^2}}{\tau_p}t\right) + \frac{\zeta^2}{\sqrt{1-\zeta^2}}\sin\left(\frac{\sqrt{1-\zeta^2}}{\tau_p}t\right)\right]\right) \tag{3.136}$$

A typical stable, underdamped step response is shown in Fig. 3.12. Some of the key parameters that can be extracted from a step response are:

(1) **Time to First Peak, t_p**: $t_p = \pi\tau/\sqrt{1-\zeta^2}$
(2) **Overshoot, *OS***: $OS = \exp\left(-\pi\zeta/\sqrt{1-\zeta^2}\right) = \frac{a}{b}$.
(3) **Decay Ratio, *DR***: The decay ratio is the square of the overshoot ratio, *i.e.* $DR = \exp\left(-2\pi\zeta/\sqrt{1-\zeta^2}\right) = \frac{c}{a}$.
(4) **Period of Oscillation, *P***: This is the time between 2 oscillations. It is given by $P = \frac{2\pi\tau}{\sqrt{1-\zeta^2}}$.
(5) **Settling Time, t_s**: The time required for the process to remain within 2.5% of the steady-state value. The first time this occurs is called the settling time.
(6) **Rise Time, t_r**: Time required to first reach the steady-state value.

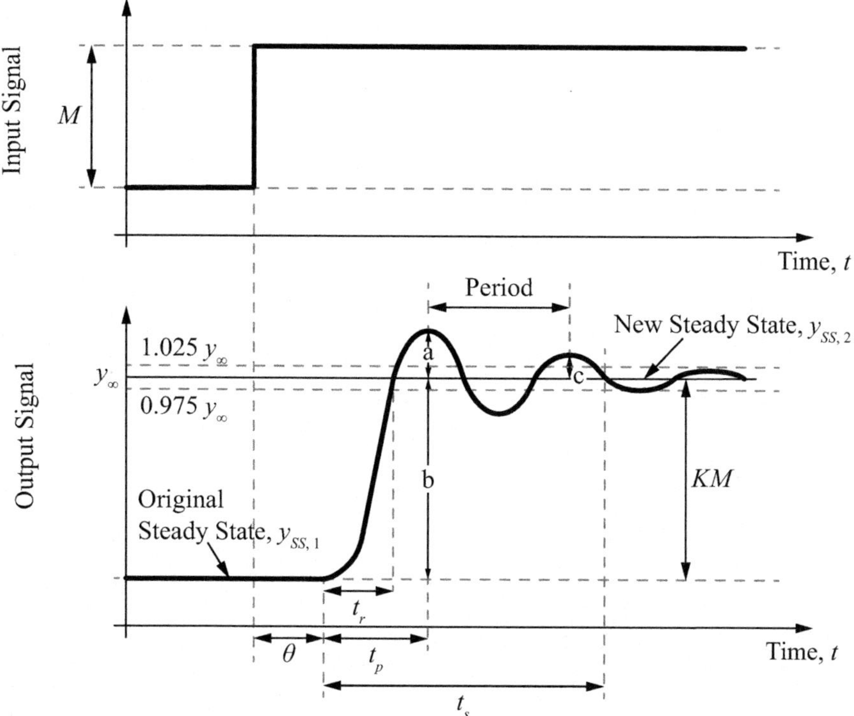

Fig. 3.12 Second-order underdamped process

For the critically damped system, where $|\zeta| = 1$, the poles of Eq. (3.132) will both be the same and contain no imaginary component. The system will be stable if the sign of $\zeta\tau_p$ is positive and unstable if $\zeta\tau_p$ is negative. The step response of this system can be written as

$$y(t) = KM\left(1 - \left(1 + \frac{t}{\tau_p}\right)e^{-\zeta t/\tau_p}\right) \tag{3.137}$$

A typical stable, critically damped step response is shown in Fig. 3.13. In general, such a system looks very similar to a first-order-plus-deadtime system and is often analysed as such. The biggest difference is the slightly slower initial response, which is often treated as a time delay.

For the **overdamped** case, where $|\zeta| > 1$, the poles of Eq. (3.132) will only be real numbers. This implies that there will be no oscillations in the step response. The system will be stable if the sign of $\zeta\tau_p$ is positive and unstable if $\zeta\tau_p$ is negative. The step response of this system can be written as[10]

[10] *Cosh* is the hyperbolic cosine function defined as $1/2(e^x + e^{-x})$ and *sinh* is the hyperbolic sine function defined as $1/2(e^x - e^{-x})$.

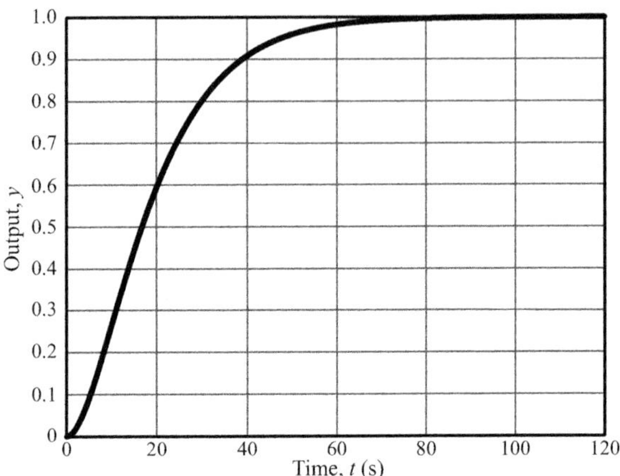

Fig. 3.13 Step response of a critically damped system ($\zeta = 1$, $\tau_p = 10$, and $K = 1$)

$$y(t) = KM\left(1 - \mathrm{e}^{-\zeta t/\tau_p}\left[\cosh\left(\frac{\sqrt{\zeta^2 - 1}}{\tau_p}t\right) + \frac{\zeta^2}{\sqrt{\zeta^2 - 1}}\sinh\left(\frac{\sqrt{\zeta^2 - 1}}{\tau_p}t\right)\right]\right) \tag{3.138}$$

A typical, stable overdamped step response is shown in Fig. 3.14. In most cases, this system can be analysed as a first-order-plus-deadtime process. It can be noted that the overdamped case has the slowest initial response of all second- and first-order systems. This slow initial response is often treated as a time delay when fitting a first-order system.

The Bode plot of a second-order system can be determined as

$$\begin{aligned}
G_{II}(j\omega) &= \frac{K}{\tau_p^2(j\omega)^2 + 2\zeta\tau_p j\omega + 1} = \frac{K}{1 - \tau_p^2\omega^2 + 2\zeta\tau_p j\omega} \\
&= \frac{K\left(1 - \tau_p^2\omega^2 - 2\zeta\tau_p j\omega\right)}{\left(1 - \tau_p^2\omega^2\right)^2 + 4\zeta^2\tau_p^2\omega^2} \\
\Rightarrow \mathrm{AR} &= \sqrt{\frac{K^2\left(1 - \tau_p^2\omega^2\right)^2 + K^2\left(-2\zeta\tau_p\omega\right)^2}{\left(\left(1 - \tau_p^2\omega^2\right)^2 + 4\zeta^2\tau_p^2\omega^2\right)^2}} = \frac{|K|}{\sqrt{\left(1 - \tau_p^2\omega^2\right)^2 + 4\zeta^2\tau_p^2\omega^2}} \\
\phi &= \arctan\left(\frac{-K2\zeta\tau_p\omega}{K\left(1 - \tau_p^2\omega^2\right)}\right) = \begin{cases}
-\arctan\frac{2\zeta\tau_p\omega}{1-\tau_p^2\omega^2} & K > 0, \left|\tau_p\omega\right| < 1 \\
-\arctan\frac{2\zeta\tau_p\omega}{1-\tau_p^2\omega^2} - 180^\circ & K > 0, \left|\tau_p\omega\right| > 1 \\
180^\circ - \arctan\frac{2\zeta\tau_p\omega}{1-\tau_p^2\omega^2} & K < 0, \left|\tau_p\omega\right| < 1 \\
-\arctan\frac{2\zeta\tau_p\omega}{1-\tau_p^2\omega^2} & K < 0, \left|\tau_p\omega\right| > 1
\end{cases}
\end{aligned} \tag{3.139}$$

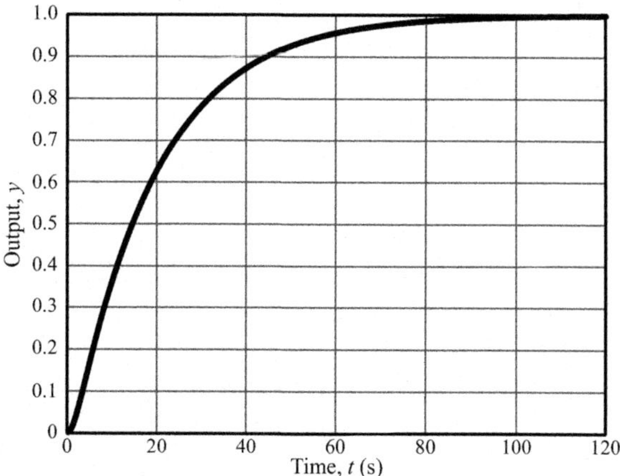

Fig. 3.14 Step response of an overdamped system ($\zeta = 2$, $\tau_p = 5$, and $K = 1$)

When dealing with second-order systems, there is a potential that the amplitude ratio can be greater than the original starting value of $|K|$. Specifically, this occurs whenever the denominator of the amplitude ratio is less than 1. Taking the denominator of the amplitude ratio for a second-order system and solving it gives

$$\begin{aligned} \left(1-\tau_p^2\omega^2\right)^2 + 4\zeta^2\tau_p^2\omega^2 &< 1 \\ 1 - 2\tau_p^2\omega^2 + \tau_p^4\omega^4 + 4\zeta^2\tau_p^2\omega^2 &< 1 \\ -2 + \tau_p^2\omega^2 + 4\zeta^2 &< 0 \\ \omega^2 &< \frac{2-4\zeta^2}{\tau_p^2} \\ |\omega| &< \sqrt{\frac{2-4\zeta^2}{\tau_p^2}} = \frac{\sqrt{2-4\zeta^2}}{|\tau_p|} \end{aligned} \tag{3.140}$$

Since the frequency is only a positive real number, the term under the square root must also only be positive (since a negative would give a complex number), that is,

$$\begin{aligned} 2 - 4\zeta^2 &\geq 0 \\ |\zeta| &\geq \sqrt{0.5} \approx 0.707 \end{aligned} \tag{3.141}$$

If the damping coefficient lies within the region given by Eq. (3.141), then there will be a hump in the Bode plot, as shown in Fig. 3.15 (top). Since these frequencies magnify the system response, it is important that special care be taken when designing controllers for such systems.

If a lead term has been added to a second-order system, as given by Eq. (3.133), then the system will display an inverse response if the zeros of the transfer function are positive, for example, as shown in Fig. 3.16. The Bode plot for this composite system can be obtained by invoking the rules for combinations of transfer functions, that is,

$$\begin{gathered}G_{II} = \frac{K(\tau_L s+1)}{\left(\tau_p^2 s^2+2\zeta\tau_p s+1\right)} = G_{II}G_L \\ \mathrm{AR} = \mathrm{AR}_{II}\mathrm{AR}_L = \frac{|K|\sqrt{1+\tau_L^2\omega^2}}{\sqrt{\left(1-\tau_p^2\omega^2\right)^2+4\zeta^2\tau_p^2\omega^2}} \\ \phi = \phi_{II}+\phi_L = \arctan\tau_L\omega - \arctan\frac{2\zeta\tau_p\omega}{1-\tau_p^2\omega^2} \\ \text{(ignoring all angle complications)}\end{gathered} \tag{3.142}$$

An example is shown in Fig. 3.17.

Representative Bode plots are shown in Fig. 3.15 for different combinations of the parameters. The properties of the second-order system are summarised in Table 3.14.

Higher-Order Systems

Analysing higher-order systems is based on the results obtained from the analysis of the simpler systems. The following are some key points to consider for the analysis in the time domain:

(1) **Poles**: The poles of the process transfer function determine the stability of the system. Table 3.6 summarises the requirements for stability.
(2) **Zeros**: If the zeros of the continuous-time transfer function are positive, then there will be an inverse response in the system.
(3) **Time Constant**: If the system is stable, then the largest stable pole in absolute magnitude can be treated as the dominant time constant of the process.

In the frequency domain, the Bode plots are obtained by combining the appropriate simple transfer functions, using the rules for composition of the amplitude ratio (multiplication) and phase angle (addition).

Example 3.14: Sketching the Expected Time-Domain Response
For the following three transfer functions, sketch the expected time-domain response of the system when a positive step change is made:

(1) $G_1 = \frac{1.54(-5s+1)}{(5s+1)(4s+2)(2s+1)}e^{-10s}$
(2) $G_2 = \frac{1.54(5s+1)}{100s^2-200s+1}$
(3) $G_3 = \frac{1.54}{\left(-100s^2+10s-1\right)(4s+1)}$

(continued on p. 91)

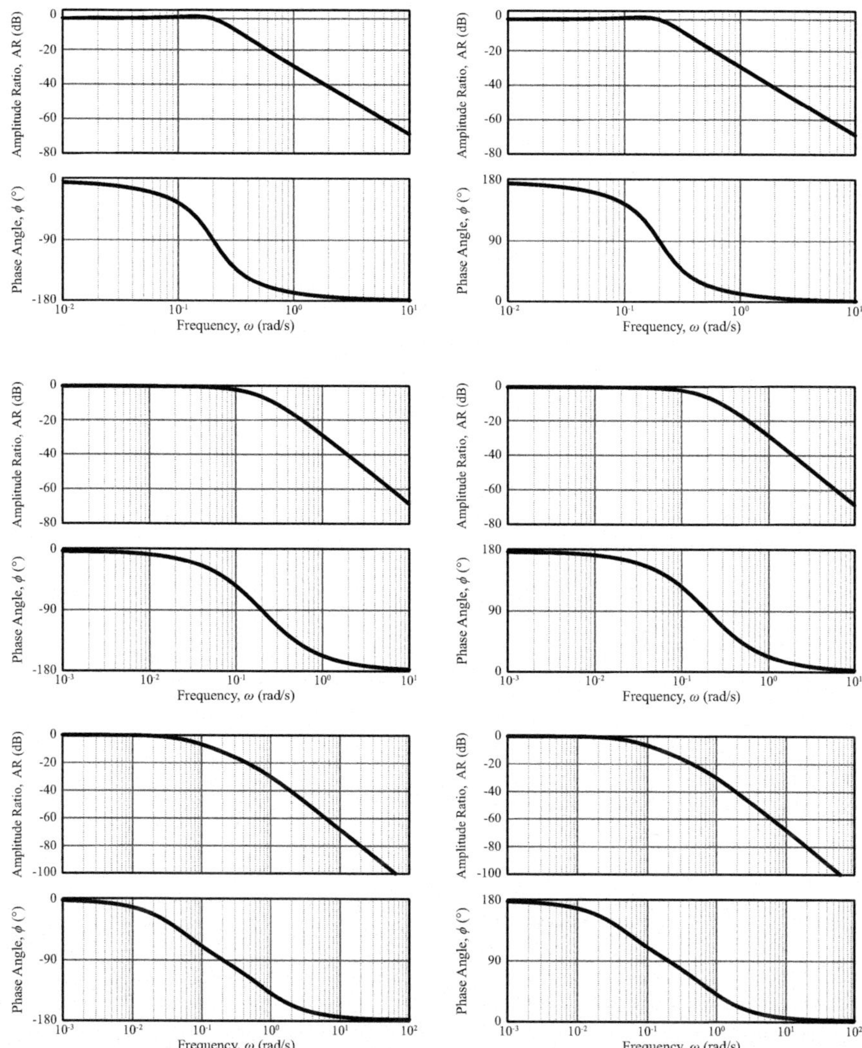

Fig. 3.15 Bode plots for $\tau_p = 5$, (top) $\zeta = 0.5$, (middle) $\zeta = 1$, (bottom) $\zeta = 2$; (left) $K = 1$ and (right) $K = -1$

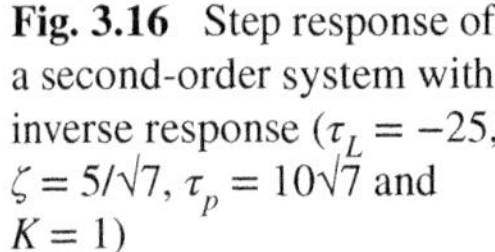

Fig. 3.16 Step response of a second-order system with inverse response ($\tau_L = -25$, $\zeta = 5/\sqrt{7}$, $\tau_p = 10\sqrt{7}$ and $K = 1$)

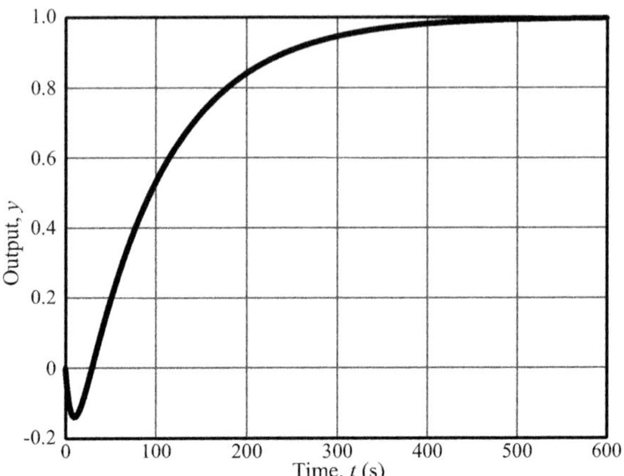

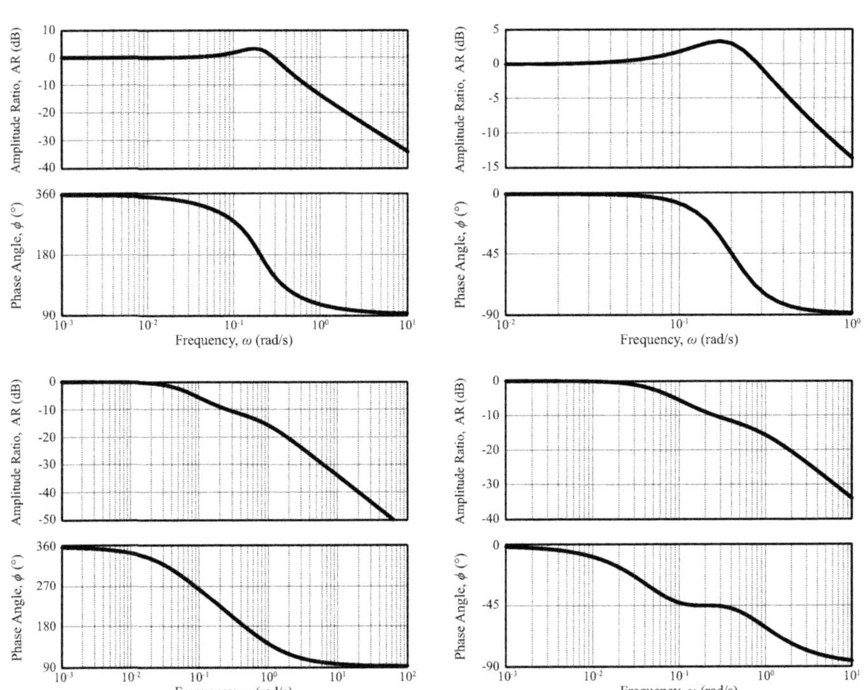

Fig. 3.17 Bode plots for (left) $\tau_L = -5$, (right) $\tau_L = 5$, and (top) $\zeta = 0.5$ and (bottom) $\zeta = 2$

Table 3.14 Basic information about a second-order system

Property		Value
Laplace transform		$G_{II} = \frac{K}{(\tau_p^2 s^2 + 2\zeta\tau_p s + 1)}$
Step response (time domain)		$\lvert\zeta\rvert < 1$: Eq. (3.136) $\lvert\zeta\rvert = 1$: Eq. (3.137) $\lvert\zeta\rvert > 1$: Eq. (3.138)
Bode plots	AR	$\frac{\lvert K\rvert}{\sqrt{(1-\tau_p^2\omega^2)^2 + 4\zeta^2\tau_p^2\omega^2}}$
	ϕ	$-\arctan\frac{2\zeta\tau_p\omega}{1-\tau_p^2\omega^2}$, see Eq. (3.139) for greater detail
Stable		Yes, if $\zeta\tau_p > 0$
Comment		This transfer function is commonly used to describe oscillations and inverse responses in a system

Solution

First Transfer Function

For the first transfer function, the gain is 1.54 / 2 = 0.77 (by computing $G_1(0)$), the zero is 0.2, the poles are − 0.2 (= − 1/5), − 0.5 (= − 2/4), and −0.5 (= −1/2), and the time delay is 10. Since the zero is positive, we expect an inverse response, while all poles are negative, which implies that the system is stable. As well, none of the poles have an imaginary component and so there are no oscillations. The plot would therefore be as given in Fig. 3.18 (left).

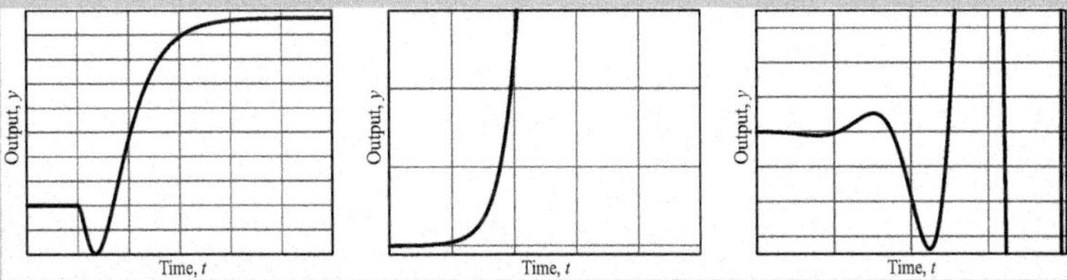

Fig. 3.18 Sketch of the transfer-function step responses: (left) First transfer function, (middle) Second transfer function, and (right) Third transfer function

Second Transfer Function

For the second transfer function, the zero is −0.2. Instead of computing the poles directly, we can note that this is a second-order system. Therefore, writing it into the standard form for a second-order system given by Eq. (3.134), it can be seen that $\tau_p = 10$ and $\zeta = -10$. Since $\tau_p\zeta < 0$, the system is unstable and without any oscillations ($|\zeta| > 1$). The "gain" is positive, which implies that the function will go towards $+\infty$. Thus, the plot would be as given in Fig. 3.18 (middle).

Third Transfer Function

For the third transfer function, there are no zeros or time delay. Similar to the second transfer function, let us rewrite the denominator into the required form to give

$$G_3 = \frac{1.54}{(-100s^2+10s-1)(4s+1)}\frac{-1}{-1} = \frac{-1.54}{(100s^2-10s+1)(4s+1)}.$$

By comparison, we can see that the second-order term has the following characteristics, $\tau_p = 10$ and $\zeta = -0.5$. This implies that the poles are unstable and oscillatory. The second term will give a stable pole ($-0.25 = -¼$). Therefore, the plot of this transfer function will be given as Fig. 3.18 (right). It can be noted that when sketching the transfer function, it is important to only give the general characteristics, while the exact values can be ignored.

Summary of the Functional Behaviour in Continuous- and Discrete-Time Domains

Table 3.15 summarises the relationship between the location of the poles of the transfer function and the generalised behaviour in both the continuous- and discrete-time domains. The *ringing* cases can only occur in the discrete-time domain. They depend on how the discretisation of the continuous-time model is performed.

Example 3.15: Origin of Ringing in Discrete-Time Systems
Consider the continuous exponential and cosine system

$$y(t) = e^{at}\cos(\omega t) \tag{3.143}$$

that is sampled with a sampling time of T_s to give a discrete-time system of the form

$$y_k = e^{akT_s}\cos(\omega kT_s) \tag{3.144}$$

which, from Table 3.3, has the z-transform of

$$\frac{1 - e^{aT_s}\cos(\omega T_s)z^{-1}}{1 - 2e^{aT_s}\cos(\omega T_s)z^{-1} + e^{2aT_s}z^{-2}} \tag{3.145}$$

We will assume that $a \in \mathbb{R}$ so that values of $a \geq 0$ will lead to an unstable system. It can be noted that the results for sine will be the same *mutatis mutandi*. Examine the effect of the sampling time on the resulting discrete-time function and its z-transform.

Solution

The behaviour of the transfer function is determined by the values of its poles, that is, the roots of the denominator. Using the quadratic formula gives the general form of the poles as

$$\begin{aligned} z &= \frac{2\psi\cos(\omega T_s) \pm \sqrt{4\psi^2\cos^2(\omega T_s) - 4\psi^2}}{2} \\ &= \psi\cos(\omega T_s) \pm \psi\sqrt{\cos^2(\omega T_S) - 1} \end{aligned} \tag{3.146}$$

where to simplify notation $\psi = e^{aT_s}$. We can note that ψ will always be positive since the exponential of any real number is positive.

In order to understand the behaviour of the discrete function, we will need to examine the determinant of the equation (the part inside the square root) and consider three cases: when the determinant is greater than zero, exactly zero, and less than zero.

Case 1: Determinant greater than zero
For the determinant to be greater than zero, it follows that

$$\cos^2(\omega T_S) - 1 > 0 \Rightarrow \cos^2(\omega T_s) > 1 \tag{3.147}$$

However, the cosine function is never greater than 1. Thus, this situation cannot occur and there will not be two distinct real roots.

Case 2: Determinant equal to zero
For the determinant to be exactly zero, it follows that

$$\cos^2(\omega T_S) - 1 = 0 \Rightarrow \cos^2(\omega T_s) = 1 \tag{3.148}$$

Taking the square root of the right-hand side and noting that there are two solutions gives

$$\cos(\omega T_s) = \pm 1 \tag{3.149}$$

which occurs when

$$\omega T_s = \pi n.$$

for $n \in \mathbb{Z}$. If we restrict ourselves to the domain $[0, 2\pi[$, we see that we have two solutions: at $n = 0$ with a value of 1 and $n = 1$ with a value of -1. In such a case, the double root will be located on the x-axis at the location of ψ for n even and at the location of $-\psi$ for n odd. When n is odd, the poles will lie in the left-hand plane of the discrete domain, which implies that we will have ringing behaviour. Stability will be determined by the value of ψ. However, there will not be any oscillatory behaviour. The general form of the discrete function is then

$$y_k = \psi^k \cos(k\pi n) \tag{3.150}$$

When n is odd, the value of $\cos(k\pi n)$ will oscillate between 1 and −1 which will give the characteristic ringing behaviour.

Case 3: Determinant less than zero
For the determinant to be less than zero, it follows that

$$\cos^2(\omega T_S) - 1 < 0 \Rightarrow \cos^2(\omega T_s) < 1 \tag{3.151}$$

In such cases, there will be two imaginary roots that are complements of each other, that is, $\psi + \gamma i$ and $\psi - \gamma i$, where γ is equal to $\psi\sqrt{1-\cos^2(\omega T_S)}$. This implies that there will be a clear oscillatory behaviour. If we restrict ourselves to the domain $[0, 2\pi[$, we can note that $\cos(\omega T_s)$ will be positive in the region $[0, 0.5\pi[\cup]1.5\pi, 2\pi[$ and negative in the region $]0.5\pi, 1.5\pi[$. The function will be exactly zero at 0.5π and 1.5π. In the positive region, this will place the poles in the right-hand side of the discrete system, while when negative it will place them in the left-hand side leading to ringing. When $\cos(\omega T_s)$ is exactly zero, we will have two roots at the origin of the system and we will have no information about the system. This brings us directly to the constraint given by the Shannon sampling theorem that the sampling time must be greater than $2/\omega$.

Table 3.15 Graphical representation of the different types of functions (The *ringing* cases can only occur in the discrete domain)

Case	Location of the poles		Unit-step response	
	Continuous	Discrete	Continuous	Discrete
Stable, exponential decay	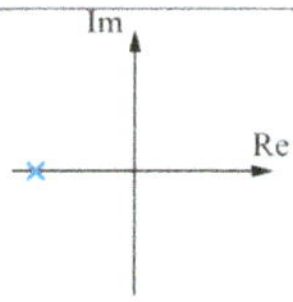 Im, Re	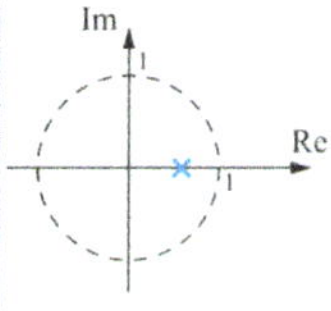 Im, 1, Re, 1		
Stable, oscillatory	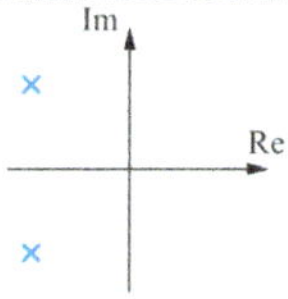 Im, Re	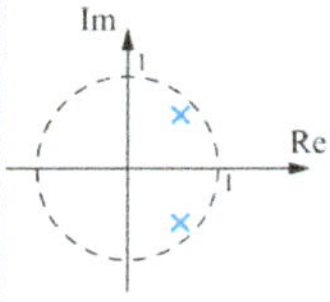 Im, 1, Re, 1		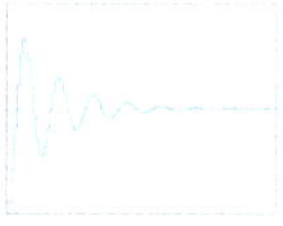
Unstable, pure oscillatory	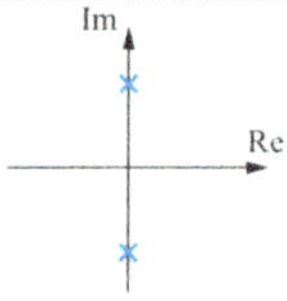 Im, Re	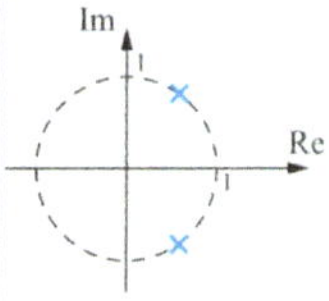 Im, 1, Re, 1	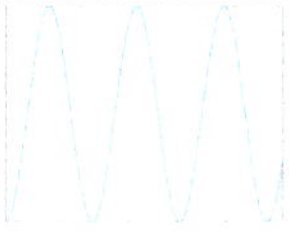	
Unstable, integrator	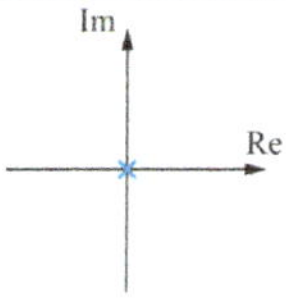 Im, Re	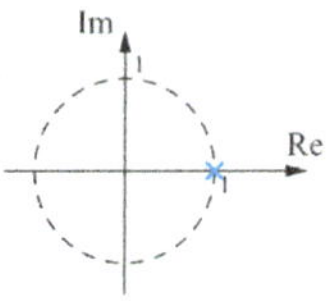 Im, 1, Re, 1		

(continued)

Table 3.15 (continued)

Case	Location of the poles		Unit-step response	
	Continuous	Discrete	Continuous	Discrete
Unstable, exponential growth	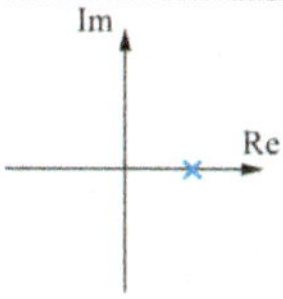 Im Re	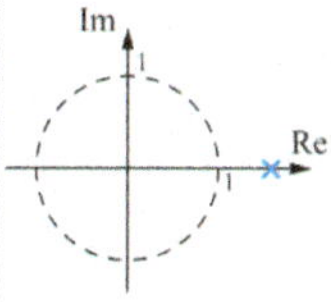 Im 1 Re 1	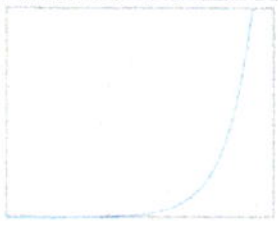	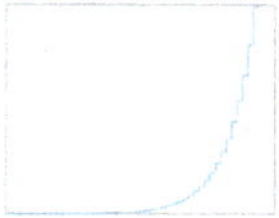
Unstable, oscillatory	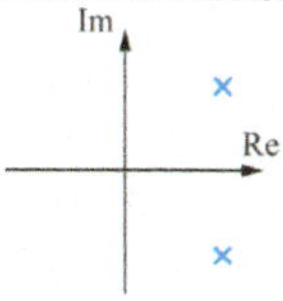 Im Re	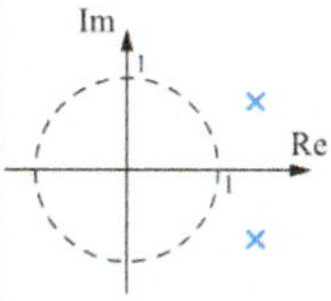 Im 1 Re 1	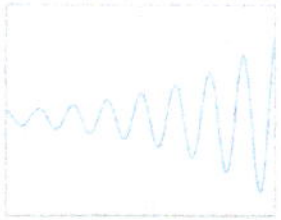	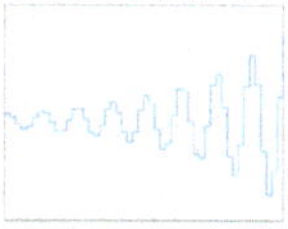
Ringing, stable		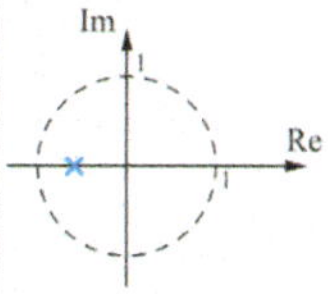 Im 1 Re 1		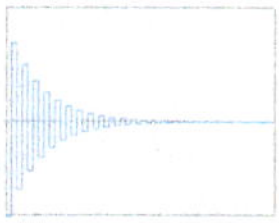
Ringing, stable oscillatory		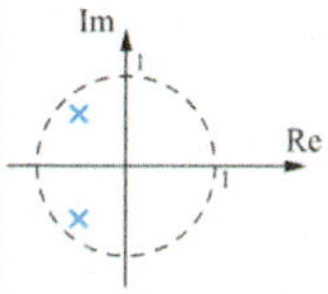 Im 1 Re 1		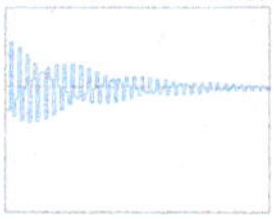
Ringing, integrator		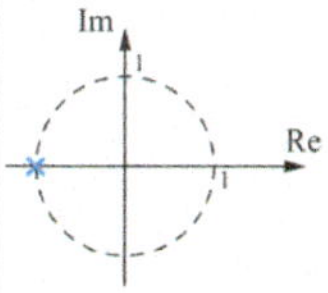 Im 1 Re 1		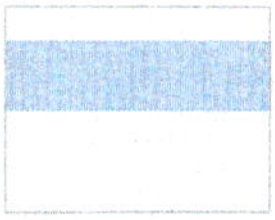
Ringing, unstable		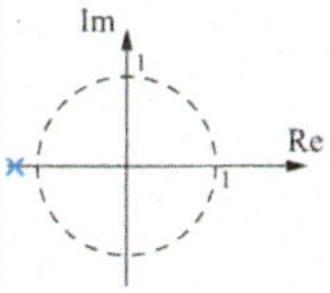 Im 1 Re 1		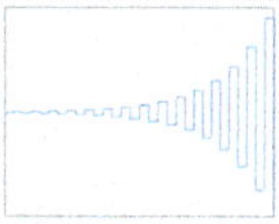
Ringing, unstable, oscillatory		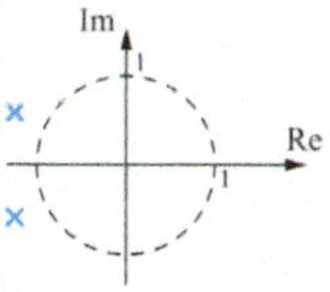 Im 1 Re 1		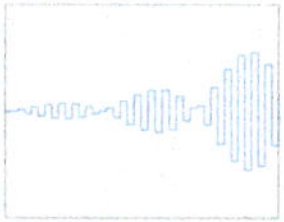
Deadbeat response		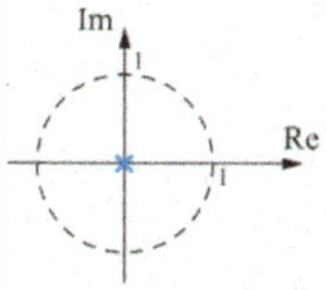 Im 1 Re 1		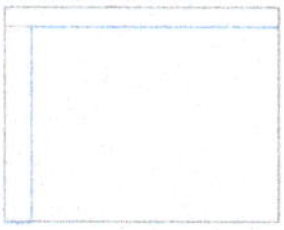

3.4 Event-Based Representations

In event-based systems, where the changes in events drive the process, a different type of model is required, where the effect of the events on the system can be more clearly seen. An **automaton**, or in the older literature machine, is a way of representing how a process changes from state to state based on discrete events. At each state, a series of inputs is recognised, which causes the automaton to move to another (or perhaps even, the same) state.

Before considering a formal definition of an automaton, let us examine a simple case and see how it could be modelled as an automaton. Consider a system that consists of only two inputs a and b. The objective of this system is to determine when the sequence of inputs is *baba*. In automaton theory, the acceptable inputs are called the **alphabet**, while the actual sequence of inputs is called a **word**. Thus, in this example, the alphabet is the set $\{a, b\}$ and the word of interest is *baba*. A **state** is defined as an internal representation of the system and its expected behaviour. In this example, we can define 5 states: *Z*0 that represents the initial state, *Z*1 when we have received the first b, *Z*2, when we have received *ba*, *Z*3, when we have received *bab*, and *Z*4, when we have received *baba*. Based on the inputs and current state, a **transition function** shows how the system will move to the next state. For example, if we are in state *Z*0 and receive an a, then the system will transition to *Z*0 (remain in the same place), while if a b is received, the system will transition to *Z*1. States can be classified as either **accepting** (also called **marked states**) or **rejecting**. An accepting state is one that has the desired outcome. In this particular example, it would be state *Z*4. A rejecting state is one that does not have the desired outcome. In this particular example, it would be all the other states. When an automaton has finished reading a word, we can note the final state of the automaton. If it is an accepting state, then we can say that the automaton accepts the word; otherwise it rejects the word. The set of all words accepted by the automaton is called the **language recognised** (or **marked**) **by the automaton**.

Mathematically, for a finite-state automaton $\mathcal{A}$, we can write this as the quintuple

$$\mathcal{A} \equiv (\mathbb{X}, \Sigma, f, x_0, \mathbb{X}_m) \tag{3.152}$$

where $\mathbb{X}$ is the finite set of states, Σ the finite set of symbols, called the alphabet of the automaton, $x_0 \in \mathbb{X}$ the initial state, $\mathbb{X}_m \subseteq \mathbb{X}$ the set of all accepting states, and f the transition function defined as

$$f : \mathbb{X} \times \Sigma \rightarrow \mathbb{X} \tag{3.153}$$

The alphabet represents all possible values that the automaton will recognise. It can consist of letters, values, or any other acceptable symbol. An automaton **generates** a language, L, which simply consists of all possible strings (irrespective of the final state) generated by the automaton. An automaton is said to **recognise** (or **mark**) a specific language L_m.

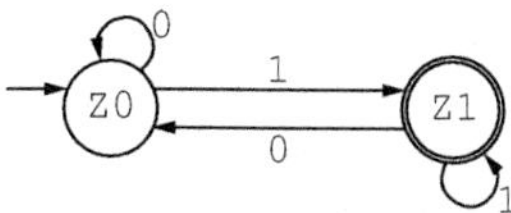

Fig. 3.19 Graphical representation of an automaton

As well, it is possible to define an output function, g, that determines what the output from the automaton will be. In general, it is defined as

$$g : \mathbb{X} \times \Sigma \to Y \tag{3.154}$$

that is, the output function depends on the state and inputs. In such cases, this automaton is called a **Mealy automaton**, which is named after George H. Mealy (1927–2010). On the other hand, when the output function, g, is defined as

$$g : \mathbb{X} \to Y \tag{3.155}$$

that is, the output function only depends on the state, then we have a **Moore automaton**, which is named after Edward F. Moore (1925–2003).

It is also possible to provide a graphical representation of an automaton. Figure 3.19 shows a typical representation of an automaton. The most important components are:

- Each state is enclosed by a circle and given a specific name.
- The initial state is shown by an arrow pointing to the state and without any additional markings.
- The transitions between the states are shown with arrows.
- On the arrows, the combination of inputs and outputs that lead to the next state are shown.
- Accepting states are enclosed by a double circle.

The sending and receiving of symbols for the inputs and outputs is assumed to be instantaneous. A finite-state automaton must allocate for each input symbol at each time point a valid transition. Often such a transition can be a **self-loop**, that is, a transition where the initial and final states are the same.

Example 3.16: Automaton for a Process
Consider the previously mentioned process where it was desired to find the word *baba*. Draw the automaton for this process and define all the components in the mathematical description of the automaton.

Solution

The automaton is shown in Fig. 3.20.

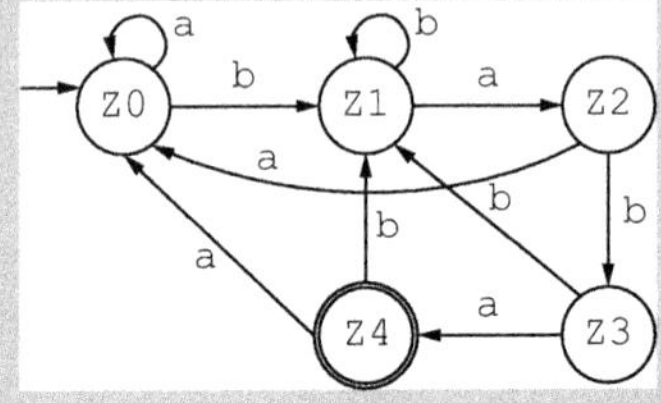

Fig. 3.20 Automaton for the example

The mathematical description can be stated as:

$$\mathbb{X} = \{Z0, Z1, Z2, Z3, Z4\}$$

$$\Sigma = \{a, b\}$$

$$x_0 = Z0$$

$$\mathbb{X}_m = \{Z4\}$$

The transition function, f, is defined as follows:

$$f(Z0, a) = Z0 \text{ (which is a self-loop)}$$

$$f(Z0, b) = Z1$$
$$f(Z1, a) = Z2$$
$$f(Z1, b) = Z1$$
$$f(Z2, a) = Z0$$
$$f(Z2, b) = Z3$$
$$f(Z3, a) = Z4$$
$$f(Z3, b) = Z1$$
$$f(Z4, a) = Z0$$
$$f(Z4, b) = Z1$$

Note that the transition function should match the arrows drawn in the schematic for the automaton!

Transitions in an automaton can be **spontaneous**, that is, we do not know exactly when a given transition occurs, for example, the transition from filling a tank to full can be spontaneous (especially, if we do not know the height of the tank). Such a transition is denoted using the symbol ε on the arrow. Spontaneous transitions often arise when there is incomplete knowledge of the system. Automata containing spontaneous transitions are said to be **nondeterministic**, since it is not possible to know in which states the automaton currently is. Another form of nondeterminism is the presence of multiple transitions with the same input, for example, two transitions from a given state with the input a leading to two different states. In such cases, it is likewise impossible to know in which state the automaton is in. In all other cases, the automaton is said to be **deterministic** since it is possible to precisely determine the next state given all past information.

The states in an automaton can be classified as follows:

(1) **Periodic States**: Periodic states are a set of states between which the automaton can oscillate. Normally, one takes the largest set of states between which the automaton can oscillate.

(2) **Ergodic States**: Ergodic states are those states once reached the automaton cannot leave.

(3) **Transient States**: All states that are not ergodic are called transient states.

3.4.1 *Analysis of Automata*

With the above mathematical model of an automaton, it is possible to analyse and manipulate the automaton. This section will briefly look at some of the ways in which we can manipulate and analyse automata.

Deadlock is said to occur if an automaton ends up in a rejecting state from which it cannot leave. Such an automaton has come to a standstill and cannot take further action. Since automata represent real processes, this is an undesirable state of events and should be avoided. A similar concept is **livelock** which is said to occur if an automaton ends up in a periodic set that consists solely of rejecting states. This means that although the automaton can make a next move it can never reach an accepting state and terminate. Both such situations are highly undesired. Together deadlock and livelock are referred to as **blocking**, since the automaton is blocked from completing its task.

Example 3.17: Blocking in an Automaton
Determine if the automaton shown in Fig. 3.21 is blocking. If it is blocking, determine what types of blocking occur.

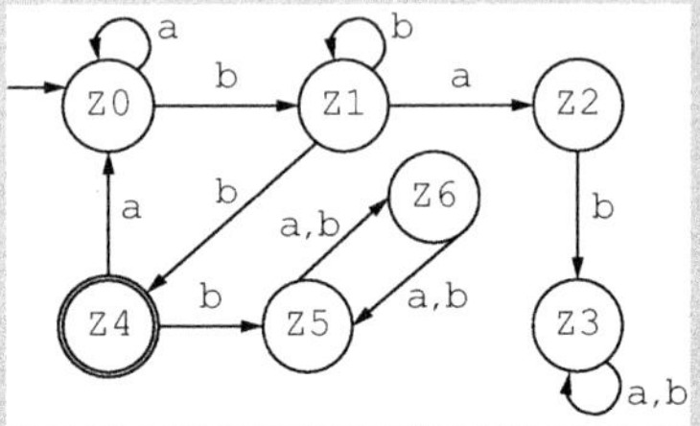

Fig. 3.21 Automaton

Solution

Looking at Fig. 3.21, state *Z3* is an ergodic state that is not accepting. Therefore, deadlock will occur in this state. Thus, the system is blocking. State *Z3* is a deadlock. States *Z5* and *Z6* form a periodic set from which there is no escape. However, neither state is accepting. Therefore, livelock will occur for these two states.

It is possible to manipulate automata. The following are some common manipulations and their definitions:

(1) **Accessible Operator** (Acc): This removes all unreachable states and their associated transitions. This operation is necessary when performing more advanced manipulations on automata to clean up the resulting automaton. This operation will not change the generated or recognised languages. A state is defined as being **unreachable** if there exists no path from the initial state to the given state.
(2) **Co-accessible Operator** (CoAc): This removes any states and their associated transitions from which one cannot end up in an accepting state. An automaton where $A = \text{CoAc}(A)$ is called a co-accessible automaton and is never blocking. This operation can change the generated language but not the recognised language.
(3) **Trim Operator** (Trim): This operation creates an automaton that is both co-accessible and accessible. The order of operation is immaterial, that is, $\text{Trim}(A) = \text{CoAc}(\text{Acc}(A)) = \text{Acc}(\text{CoAc}(A))$.

Example 3.18: Trimming an Automaton
Apply the trim operation to the automaton shown in Fig. 3.21.

Solution

Quickly looking at the automaton in Fig. 3.21, we see that there are no states that cannot be accessed from the initial state. Thus, we need to consider the co-accessible operator. Here, we see that we need to remove states *Z*5, *Z*6, and *Z*3 since we know that they are blocking. However, by removing *Z*3, we now make *Z*2 blocking. Therefore, we must also remove it. In general, this process is iterative and continues until we have removed all states or there are no more states to remove. The final automation in shown in Fig. 3.22.

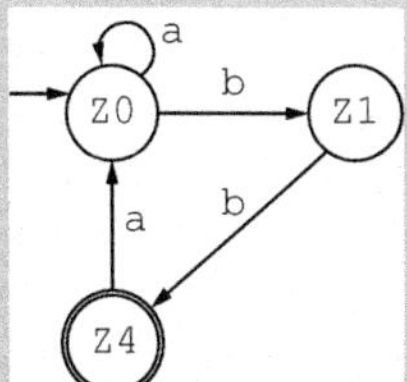

Fig. 3.22 The trimmed automaton

3.4.2 Combining Automata

Having looked at various operations on single automata, it is worthwhile to consider manipulating two or more different automata. When we wish to combine two automata, we can perform two different operations: **product**, denoted by ×, and **parallel composition**, denoted by ||. Parallel composition is often also called synchronous composition.

Product composition is defined as combination of two automaton considering only the letters of the alphabet that the two automata have in common, that is, $\Sigma_1 \cap \Sigma_2$. Formally, for two automata G_1 and G_2, their product can be written as[11]

$$G_1 \times G_2 \equiv \mathrm{Acc}(\mathbb{X}_1 \times \mathbb{X}_2, \Sigma_1 \cup \Sigma_2, f, \left(x_{0_1}, x_{0_2}\right), \mathbb{X}_{m_1} \times \mathbb{X}_{m_2}) \tag{3.156}$$

where

$$f((x_1, x_2), e) \equiv \begin{cases} (f_1(x_1, e), f_2(x_2, e)) & \text{if } e \text{ is a valid input at the given point} \\ \text{undefined} & \text{otherwise} \end{cases} \tag{3.157}$$

[11] Note that the event set here is defined as the union of both event sets, since we may wish to keep track of all the events in future cases.

In product composition, the transitions of the two automata are always synchronised on a common event. This implies that a transition occurs only if the input is a valid input for both automata. The states of $G_1 \times G_2$ are given as the pair (x_1, x_2), where x_1 is the current state of G_1 and x_2 is the current state of G_2. It follows from the definition of product composition that

$$\begin{aligned} L(G_1 \times G_2) &= L(G_1) \cap L(G_2) \\ L_m(G_1 \times G_2) &= L_m(G_1) \cap L_m(G_2) \end{aligned} \tag{3.158}$$

Product composition has the following properties:

(1) It is commutative up to a re-ordering of the state components in the composed states.
(2) It is associative. This implies that $G_1 \times G_2 \times G_3 \equiv (G_1 \times G_2) \times G_3 = G_1 \times (G_2 \times G_3)$.

Example 3.19: Product of Two Automata
Consider the automata shown in Figs. 3.23 and 3.24. Determine the product of these two automata.

Solution

Before we can start with drawing the final automaton, it makes sense to first consider which inputs are valid for the final automaton. From Fig. 3.23, it can be seen that, for G_1, the inputs are $\{a, b, g\}$. Similarly, from Fig. 3.24, it can be seen that, for G_2, the inputs are $\{a, b\}$. Therefore, the common set is $\{a, b\}$.

Fig. 3.23 G_1

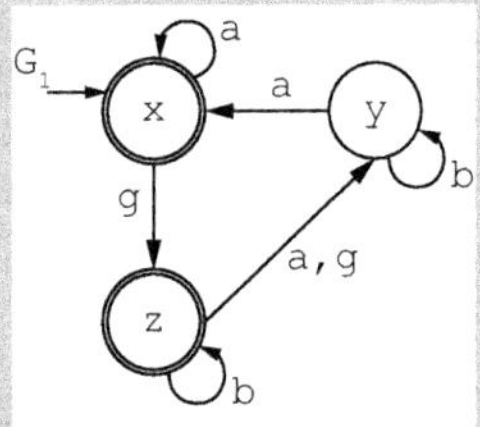

Fig. 3.24 G_2

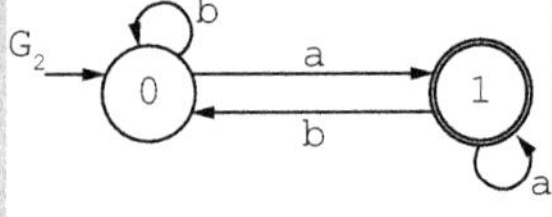

When drawing the final automaton, it helps to start from the initial states of both and work through all the possible transitions and draw the next state. Thus, starting from $(x, 0)$, with an input of a, G_1 remains in state x, but G_2 goes to state 1. Therefore, the new state in the product automaton will be $(x, 1)$. This is the only valid transition since from $(x, 0)$, an input of b is not valid for G_1. An input must be valid for both automata for it to occur in the product. Note that g will not occur since it is not a common input. We will then continue in the same fashion by looking at the possible states and how the inputs would affect them. A state will be marked if all the states in the original automaton are marked. The final automaton is shown in Fig. 3.25.

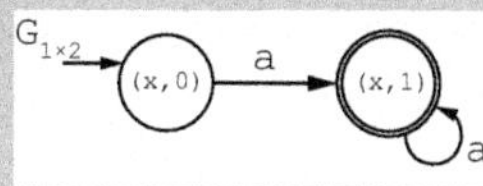

Fig. 3.25 The product of G_1 and G_2

As can be seen, product composition is relatively restrictive in that a given input must be valid for both automata. One way to relax this constraint is to consider parallel composition. In parallel composition, the inputs are split into two parts: **common inputs** and **private inputs**. A private input only pertains to a given automaton, while common inputs are shared with other automata. Formally, the parallel composition of two automata G_1 and G_2, denoted as $G_1 \parallel G_2$, is defined as

$$G_1||G_2 \equiv \text{Acc}(\mathbb{X}_1 \times \mathbb{X}_2, \Sigma_1 \cup \Sigma_2, f, (x_{0_1}, x_{0_2}), \mathbb{X}_{m_1} \times \mathbb{X}_{m_2}) \tag{3.159}$$

where

$$f((x_1, x_2), e) \equiv \begin{cases} (f_1(x_1, e), f_2(x_2, e)) & \text{if } e \text{ is a valid common input for both automata} \\ (f_1(x_1, e), x_2) & \text{if } e \text{ is a private event to } G_1 \\ (x_1, f_2(x_2, e)) & \text{if } e \text{ is a private event to } G_2 \\ \text{undefined} & \text{otherwise} \end{cases} \tag{3.160}$$

Note that the only difference between parallel and product composition is how the transition function is defined.

Product composition has the following properties:

(1) It is commutative up to a re-ordering of the state components in the composed states.

(2) It is associative. This implies that $G_1 \parallel G_2 \parallel G_3 \equiv (G_1 \parallel G_2) \parallel G_3 = G_1 \parallel (G_2 \parallel G_3)$.

Example 3.20: Parallel Composition of Two Automata
Consider the same two automata as for Example 3.19. Determine the parallel composition of these two automata.

Solution

The general procedure for solving such a problem is similar to that of the product composition. First, we need to determine which inputs belong to which categories. Since input *g* only affects G_1, it is a private input for G_1. The other two inputs $\{a, b\}$ are the common inputs to both automata.

Again, we start with the initial states of both automata and work our way through. From the initial state of $(x, 0)$, we have two valid inputs (the common input *a* and the private input *g*). The common input *a* will bring us as before to the state $(x, 1)$, while the private input *g* will bring us to the state $(z, 0)$. In the state $(z, 0)$, there are three valid inputs (the common events *a* and *b* as well as the private input *g*). Input *a* will bring us to state $(x, 1)$, while input *b* will be a self-loop. Private input *g*, which only affects G_1, will bring us to state $(y, 0)$. The final automaton is shown in Fig. 3.26.

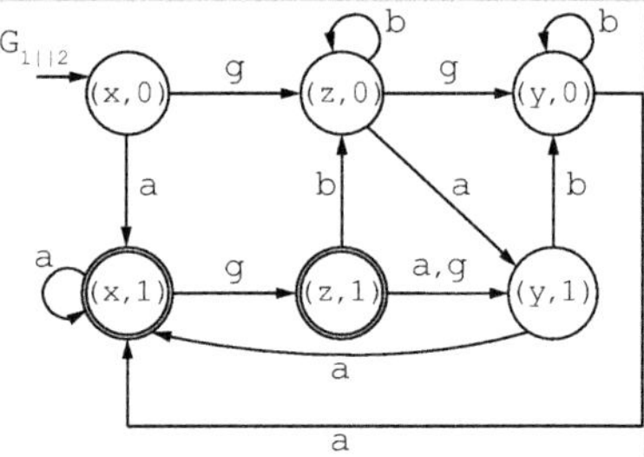

Fig. 3.26 Parallel composition of G_1 and G_2

3.4.3 *Timed Automata*

A **timed automaton** is an automaton that is linked with a clock. A timed automaton has four components: the finite-state automaton, the **clock**, the **invariants**, and the **guards**. Such automata can model minimal and maximal times associated with a given action. Figure 3.27 shows part of a timed automaton. The clock variable *c* represents the elapsed time since the last reset, which is denoted as $c \equiv 0$. A guard shows when a transition is activated and can be implemented, for example, in Fig. 3.27, we see that the transition $Z0 \rightarrow Z1$ can only be considered once $c \geq 180$ s. An invariant shows how long the automaton can remain in a given state, for example, in Fig. 3.27, we see that the process can remain in state *Z0* until

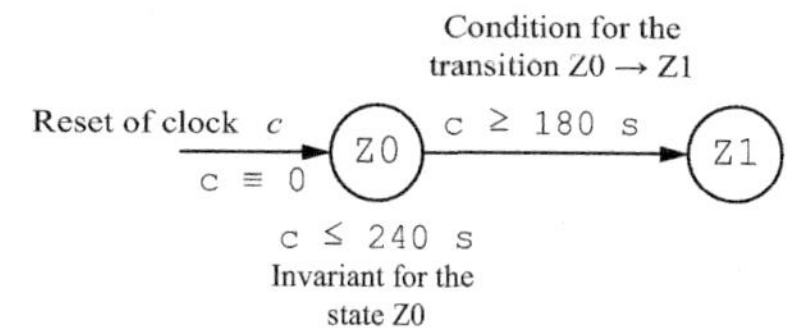

Fig. 3.27 Timed automaton

$c = 240$ s, at which point it must leave the state. The implies that a time automaton has an infinite number of possible realisations, since the system can switch states at any point between 180 and 240 s. Thus, a timed automaton has an infinite state space.

Mathematically, a timed automaton can be represented as the septuple

$$\mathcal{A} \equiv (\mathbb{X}, \Sigma, f, \mathbb{C}, I, x_0, \mathbb{X}_m) \tag{3.161}$$

where $\mathbb{C}$ is a finite set of clock variables and I is the invariant function that links the states with the transition constraints, that is,

$$I : \mathbb{X} \rightarrow \Phi(\mathbb{C}) \tag{3.162}$$

where $\Phi(\mathbb{C})$ is the set of transition constraints δ. The transition constraints δ are always either true or false based on the current clock variables. The allowed transition constraints are (where $a \in \mathbb{R}$ and $c \in \mathbb{C}$):

- $\delta \equiv$ (c $\leq$ a)
- $\delta \equiv$ (c $=$ a)
- $\delta \equiv$ (c $\geq$ a)
- $\delta \equiv$ (δ1 OR δ2)
- $\delta \equiv \neg$(δ)
- $\delta \equiv \varnothing$ or $\{\}$

The transition function is defined as

$$f : \mathbb{X} \times \Sigma \times \Phi(\mathbb{C}) \rightarrow \mathbb{X} \times 2^{|\mathbb{C}|} \tag{3.163}$$

where $2^{|\mathbb{C}|}$ represents the power set of $\mathbb{C}$.

In certain formalisms, transitions can be specified as either **urgent** or **nonurgent**. Urgent transitions will always be taken as soon as possible, while nonurgent transitions can wait. Normally, spontaneous transitions are assumed to be nonurgent.

3.5 Further Reading

The following are additional resources on the topics mentioned in this chapter:

(1) **General Engineering Mathematics**: E. Kreyszig (2011). *Advanced Engineering Mathematics* (10th edition), New York, New York, USA: Wiley.
(2) **Laplace Transform and Continuous-Time Analysis**: G. Doetsch (1974). *Introduction to the Theory and Application of the Laplace Transformation*, Berlin, Germany: Springer.
(3) **Z-Transform and Discrete-Time Analysis**

   a. R. El Attar (2005). *Lecture notes on Z-Transform*, Morrisville, North Carolina, USA: Lulu Press.
   b. K. Ogata (1995). *Discrete-Time Control Systems* (Second edition), Upper Saddle River, New Jersey, USA: Prentice-Hall.

(4) **Automata**

   a. J. Carroll and D. Long (1989). *Theory of finite automata*, Englewood Cliffs, New Jersey, USA: Prentice-Hall.
   b. C. G. Cassandras and S. Lafortune (2008), *Introduction to Discrete Event Systems*, (Second edition), New York, New York, USA: Springer.

3.6 Chapter Problems

Problems at the end of the chapter consist of three different types: (a) Basic Concepts (True/False), which seek to test the reader's comprehension of the key concepts in the chapter; (b) Short Exercises, which seek to test the reader's ability to compute the required parameters for a simple data set using simple or no technological aids; and (c) Computational Exercises, which require not only a solid comprehension of the basic material, but also the use of appropriate software.

3.6.1 Basic Concepts

Determine if the following statements are true or false and state why this is the case.

(1) A process satisfying the principles of superposition and homogeneity is said to be linear.
(2) In a time-varying model, the parameters themselves vary with respect to time.

(3) In a lumped-parameter model, there are space derivatives.
(4) A noncausal system depends on future values.
(5) A system with memory only cares about the current value of the process.
(6) A state-space representation provides a model linking states, inputs, and outputs.
(7) A transfer function can only be written for linear processes.
(8) Every transfer function has a unique state-space representation.
(9) Prediction-error models are discrete-time models of the process.
(10) A white-noise signal depends on past values of the noise.
(11) In a Box-Jenkins model, the order of the A-polynomial is fixed to zero.
(12) In an autoregressive exogenous model, only the orders of the C- and D-polynomials are zero.
(13) It is not possible to convert a continuous model into a discrete model.
(14) A process at steady state will experience wild, unpredictable swings in values.
(15) The gain of a process represents the transient behaviour of the process.
(16) The process time constant represents the delay before a process responds.
(17) A continuous transfer function with poles of -2, -1, and 0 is stable.
(18) A continuous transfer function with poles of 1, 2, and 5 is stable.
(19) A discrete transfer function with poles of 0.5, -0.5, and 1 is unstable.
(20) A discrete transfer function with poles of 0.25, 0.36, $0.25\pm0.5i$ is stable.
(21) A continuous state-space model with eigenvalues of $0.25\pm2i$ is stable.
(22) A discrete state-space model with eigenvalues of $\pm0.25i$ is stable.
(23) The alphabet of an automaton represents the inputs allowed into the process.
(24) An accepting state is a state that has the desired outcome.
(25) The language recognised by an automaton is the set of all words accepted by the automaton.
(26) In a Mealy automaton, the output function depends only on the states.
(27) Blocking occurs when an automaton cannot reach an accepting state.
(28) We denote spontaneous transitions by ε.
(29) An automaton with spontaneous transitions is called a deterministic automaton.
(30) The co-accessible operator removes all states and their associated transitions that are unreachable from the initial state.
(31) In a timed automaton, a guard determines the maximal time in which we can remain in a given state.
(32) For a timed automaton, $\delta \equiv$ `(c ≤ 1.000)` is an acceptable time constraint.
(33) For a timed automation, $\delta \equiv$ `(c = 'abcd')` is an acceptable time constraint.
(34) There is no escape from an ergodic state.

3.6.2 Short Questions

These questions should be solved using pen and paper. Appropriate software for drawing the required diagrams can also be used to assist with the design.

(35) For the following models, classify them based on the information in this chapter. Are the models linear, time-invariant, lumped parameter, memoryless, or causal?

a. $y_{k+1} = 4y_k + 7u_{k+5} - 5e_k$.

b. $\frac{\partial^2 T}{\partial x^2} = -\alpha(t)\frac{\partial T}{\partial t}$, where α is a parameter that depends on time.

c. $y_{k+1} = -4y_k - 3e_k$

d. $\frac{\partial T}{\partial t} = -\alpha(t)u(t+1)$, where α is a parameter that depends on time.

(36) For the following continuous-time transfer functions, determine their stability. For the stable transfer functions, determine the gain, time constant, and time delay.

a. $G(s) = \frac{5(s+1)}{6s^3+11s^2+6s+1}e^{-3s}$

b. $G(s) = \frac{-5}{150s^3+65s^2+2s-1}e^{-10s}$

c. $G(s) = \frac{4.5}{15s^2+8s+1}e^{-10s}$

d. $G(s) = \frac{4.5}{15s^2+26s+7}e^{-7s}$

(37) For the following continuous-time state-space models, determine their stability. Convert the models to a transfer function. For the stable models, determine their gain and time constant.

a.
$$\frac{\mathrm{d}\vec{x}}{\mathrm{d}t} = \begin{bmatrix} 5 & 0 \\ 0 & -2 \end{bmatrix}\vec{x} + \begin{bmatrix} 2 \\ 1 \end{bmatrix}\vec{u}$$
$$y = \begin{bmatrix} 1 & 0 \\ 0 & 1 \end{bmatrix}\vec{x}$$

b.
$$\frac{\mathrm{d}\vec{x}}{\mathrm{d}t} = \begin{bmatrix} -5 & 0 \\ 0 & -2 \end{bmatrix}\vec{x} + \begin{bmatrix} -2 \\ 1 \end{bmatrix}\vec{u}$$
$$y = \begin{bmatrix} 2 & 0 \\ 0 & 1 \end{bmatrix}\vec{x}$$

c.
$$\frac{\mathrm{d}\vec{x}}{\mathrm{d}t} = \begin{bmatrix} -2 & 1 & 2 \\ 0 & -3 & 2 \\ 0 & 0 & -1 \end{bmatrix}\vec{x} + \begin{bmatrix} 1 \\ -0.5 \\ 2 \end{bmatrix}\vec{u}$$
$$y = \begin{bmatrix} 2 & 0 & 0 \\ 0 & -1 & 0 \\ 0 & 0 & 1 \end{bmatrix}\vec{x}$$

(38) For the following discrete-time models, determine their stability:

a. $y_{k+1} = 4y_k + 7u_{k+5} - 5e_k$

b. $y_{k+1} = \frac{z^{-5}}{1-4z^{-1}} u_k$

c. $y_{k+1} = \frac{z^5+z^4}{z^6+z^5+z^4+z^3+z^2+z^1+1} u_k$

d.

$$\vec{x}_{k+1} = \begin{bmatrix} -0.5 & 1 & 2 \\ 0 & 0.5 & 2 \\ 0 & 0 & 0.25 \end{bmatrix} \vec{x}_k + \begin{bmatrix} 1 \\ -0.5 \\ 2 \end{bmatrix} \vec{u}_k$$

$$\vec{y}_k = \begin{bmatrix} 2 & 0 & 0 \\ 0 & -1 & 0 \\ 0 & 0 & 1 \end{bmatrix} \vec{x}_k$$

e.

$$\vec{x}_{k+1} = \begin{bmatrix} -1.5 & 0 & 0 \\ 3 & 1.5 & 0 \\ 1 & 2 & 2.25 \end{bmatrix} \vec{x}_k + \begin{bmatrix} -1 \\ 0.5 \\ 3 \end{bmatrix} \vec{u}_k$$

$$\vec{y}_k = \begin{bmatrix} 1 & 0 & 0 \\ 0 & 1 & 0 \\ 0 & 0 & 1 \end{bmatrix} \vec{x}_k$$

(39) What is the relationship between the eigenvalues of the state-space model and the time constant (as determined from the transfer function)?

(40) Derive the compact state-space representations for addition (given by Eq. (3.71)), multiplication (given by Eq. (3.72)), the inverse (given by Eq. (3.73)), and the transpose (given by Eq. (3.74)).

(41) For the automata shown in Fig. 3.28, classify the states into marked, ergodic, periodic, and transient. Determine if there is blocking. If present, give the type of blocking encountered.

(42) Perform the trim operation on the automata in Fig. 3.28.

(43) Using the automata given in Fig. 3.29, perform the following operations: `K1×K2`, `K1||K2`, `K1×K3`, `K1||K2||K3`, and `K1×K2×K3`.

(44) Draw the automata for the following processes:

a. Given the letters *a* and *b*, find the string *abab*.
b. Given the letters *c*, *d*, and *e*, find the string *decd*.
c. Given the letters *g*, *h*, and *i*, find the strings *hig* and *high*.

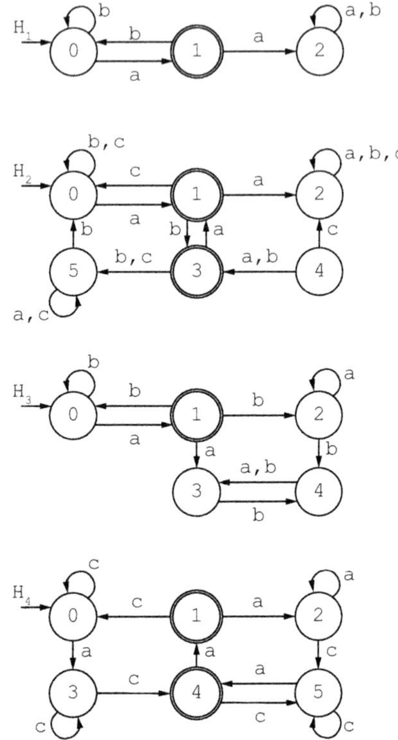

Fig. 3.28 Automata for Questions 41 and 42

3.6.3 *Computational Exercises*

The following problems should be solved with the help of a computer and appropriate software packages, such as MATLAB®.

(45) Model a system that you are familiar with. Be sure to include all the differential equations required for the process.

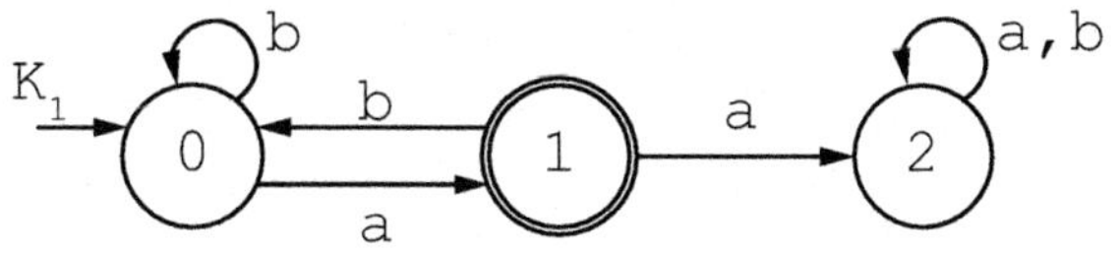

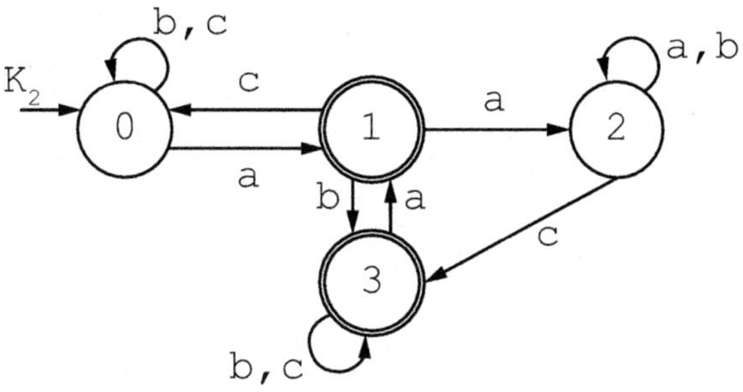

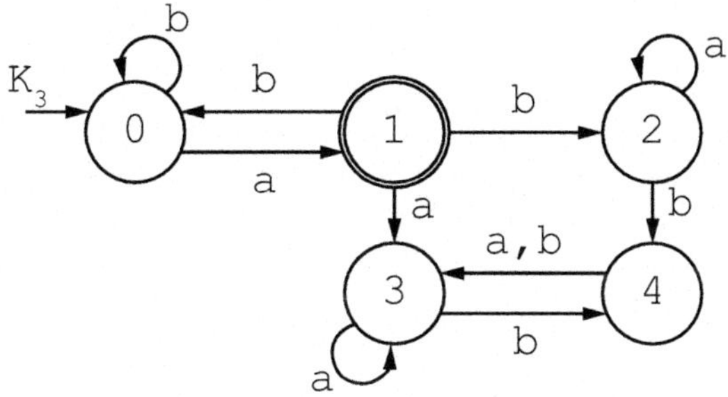

Fig. 3.29 Automata for Question 42

Chapter 4
Schematic Representation of a Process

This chapter will describe the most common schematic methods for representing a process, including **block diagrams**, **piping-and-instrumentation diagrams (P&IDs)**, **process-flow diagrams** (PFDs), and **electrical and logic circuit diagrams**.

4.1 Block Diagrams

A block diagram is an abstract way of representing a system in the frequency domain that allows for all the messy details to be hidden and only the essential elements to be shown.[1] The **basic block diagram** consists of three parts, as shown in Fig. 4.1. On the left, entering the block diagram, is the input, denoted by U, while on the right, leaving the block diagram, is the output, denoted by Y. Inside the block, the **process model**, denoted by G, is shown. In most cases, the exact form of the process model will not be specified, but it can be any relationship between the input and output.

Another common block is the **summation block**, which shows how two or more signals are to be combined. Figure 4.2 shows a typical summation block. The signs inside the circle show whether the signal coming in should be added or subtracted. Given the common nature of a summation block, two simplifications can be made. Rather than using a full circle, the signals are simply shown to connect and the signs are shown beside each signal. This is shown in the bottom

[1] Often, it will also be used for a time-domain representation by ignoring the composition and summation rules and using slightly different models. In such cases, they are better called process-flow diagrams, which are described in Sect. 4.2.

Y. A. W. Shardt, *Automation Engineering*,
https://doi.org/10.1007/978-3-031-92394-4_4

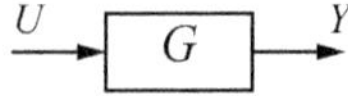

Fig. 4.1 Basic block diagram

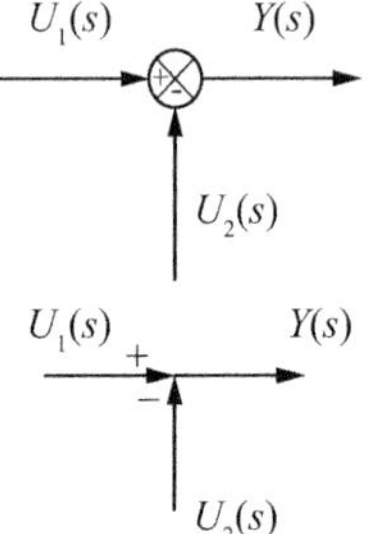

Fig. 4.2 Summation block: (top) full form and (bottom) short-hand equivalent

part of Fig. 4.2. A further simplification is to completely ignore any positive signs and only give the negative signs beside the appropriate signals. A commonly used standard is to place the summation signs on the left-hand side of the arrow.

One of the most useful features of block diagrams is the ability to easily compute the relationship between the different signals. For the basic block shown in Fig. 4.1, the relationship between the input and output can be written as:

$$Y = GU \tag{4.1}$$

while for the summation block, it can be written as

$$Y = U_1 - U_2. \tag{4.2}$$

Now, consider the process shown in Fig. 4.3, where it is desired to determine the relationship between Y and U, where there are three blocks in series. The naïve approach would be to denote each of the outputs from the two intermediate steps as Y_1 and Y_2 and then to write the relationships between each of these variables to obtain the final result, that is,

$$Y_1 = G_1 U \tag{4.3}$$

$$Y_2 = G_2 Y_1 = G_2 G_1 U \tag{4.4}$$

$$Y = G_3 Y_2 = G_3 G_2 G_1 U. \tag{4.5}$$

This shows that the final relationship can be written as $Y = G_3 G_2 G_1 U$. The easier approach is to note that since these process blocks are in series, the process models can be multiplied together to give the desired result (as the final result shows).

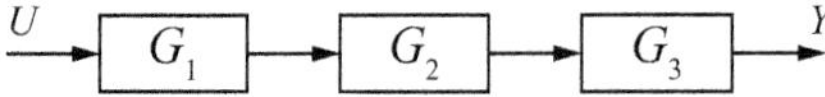

Fig. 4.3 Block-diagram algebra: In order to relate U and Y, the transfer functions between the two points need to be multiplied, thus, $Y = G_3 G_2 G_1 U$

However, the naïve approach is very useful if the system is very complex with many different summation blocks and functions.

Example 4.1: Complex Block Diagrams
Consider the closed-loop system shown in Fig. 4.4 and derive the expression for the relationship between R and Y. Assume that all signals are in the frequency domain.

Solution

Starting from the desired signal R and moving towards Y, the following relationships can be written

$$\begin{aligned}\varepsilon &= R - G_m Y\\ U &= G_c\varepsilon\\ Y &= G_p G_a U + G_d D.\end{aligned}$$

Substituting the first relationship into the second and then the combination into the third gives

$$Y = G_p G_a G_c (R - G_m Y) + G_d D.$$

Re-arranging this equation gives

$$Y = G_p\ G_a G_c R - G_p\ G_a G_c G_m Y + G_d D.$$

Solving this equation for Y gives

$$\boxed{Y = \frac{G_p G_a G_c}{1 + G_p G_a G_c G_m} R + \frac{G_d}{1 + G_p G_a G_c G_m} D}.$$

Since we are only interested in the relationship between Y and R, we can set $D = 0$ to get

$$Y = \frac{G_p G_a G_c}{1 + G_p G_a G_c G_m} R.$$

This equation is commonly encountered in closed-loop control.

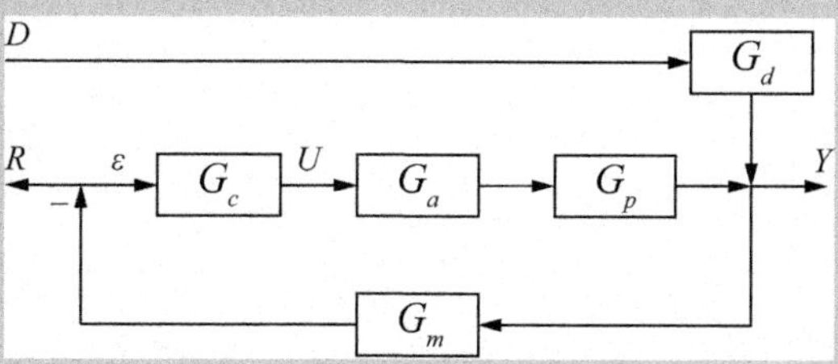

Fig. 4.4 Generic closed-loop, feedback control system

The creation of block diagram for a given process can be quite a complex task, but it is worthwhile as it allows the essential features of the system to be abstracted out and understood.

4.2 Process-Flow Diagrams

The process-flow diagram is a simplified diagram of the process, where only the key components and connections are shown. For complex processes, the process-flow diagram is often created using blocks, where the blocks show a subprocess. In such cases, they can approximate the block diagrams described in Sect. 4.1.

A process-flow diagram for a single process normally contains the following elements: process piping, key components, key valves, control valves, connections to other systems, key bypass and recycle streams, and process-flow names.

Figure 4.5 shows a typical process-flow diagram. The rules for creating a process-flow diagram are similar to the rules for creating a piping-and-instrumentation diagram, given in Sect. 4.3. The only difference is the level of detail.

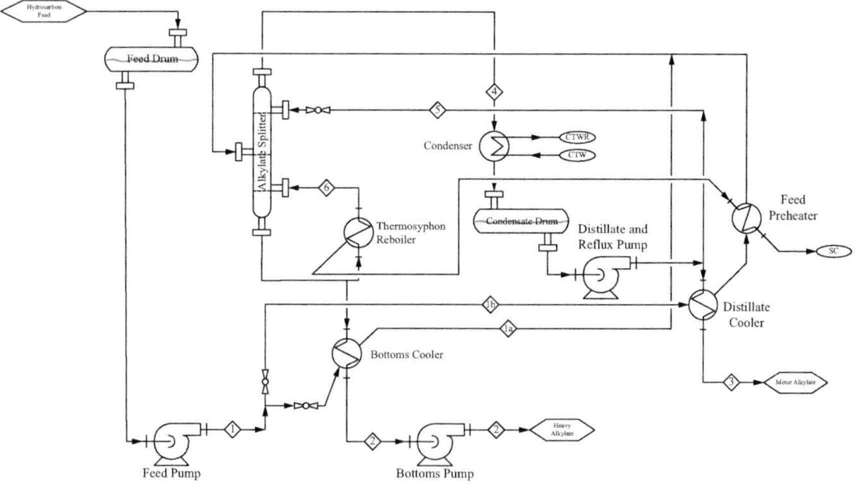

Fig. 4.5 Process-flow diagram for an alkylate splitter

4.3 Piping-and-Instrumentation Diagrams (P&ID)

The piping-and-instrumentation diagram (P&ID) is a detailed description of the process, where all connections and components are shown. A P&ID contains, in addition to the information found in a process-flow diagram, the following information:

(1) type and identification number for all components;
(2) piping, fittings with nominal diameters, pressure stages, and materials;
(3) drives; and
(4) all measurement and control devices.

Figure 4.6 shows a typical P&ID for part of a chemical plant.

4.3.1 P&ID Component Symbols According to the DIN EN 62424 Standard

Table 4.1 shows the typical component symbols according to the DIN EN 62424 Standard.[2] Additional symbols can be found in the standard itself.

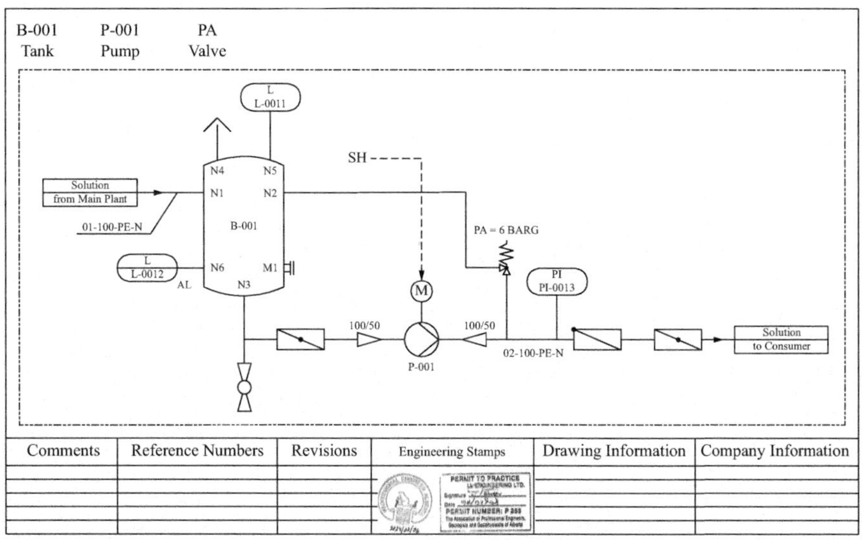

Fig. 4.6 P&ID for a gas-chilling-and-separation plant according to Canadian design standards (Note the engineering stamp in the bottom middle box.)

[2] The older German standard DIN 19227 matches closely the ISA or North American standard. Differences include how the different sensors or functions are denoted and minor details regarding the proper location for various pieces of additional information. This book will follow the new standard without necessarily making any comments about the other possibilities in order to avoid confusion.

Table 4.1 Component symbols for P&IDs according to the DIN EN 62424 Standard

Symbol	Name
	Pipe
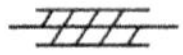	Insulated pipe
	Jacketed pipe
	Cooled or heated pipe
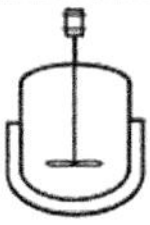	Vessel (chemical reactor) with jacket
	Horizontal pressurised vessel
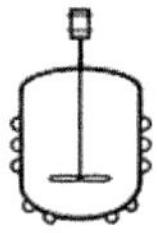	Half-pipe reactor
	Fluid-contacting column
	Pump
	Vacuum pump or compressor
	Bag
	Column with trays
	Fan
	Axial fan
	Radial fan

(continued)

Table 4.1 (continued)

Symbol	Name
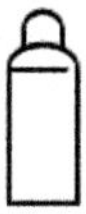	Gas bottle
	Furnace
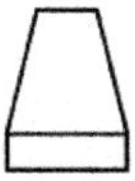	Cooling tower
	Dryer
	Cooler
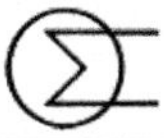	Heat exchanger without cross of fluxes
	Heat exchanger with cross of fluxes
	Plate heat exchanger
	Spiral heat exchanger
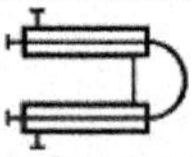	Double-pipe heat exchanger
	Fixed straight-tube heat exchanger
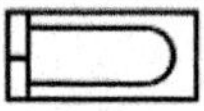	U-shaped-tube heat exchanger
	Finned-tube heat exchanger with axial fan
	Covered gas vent

(continued)

Table 4.1 (continued)

Symbol	Name
	Curved gas vent
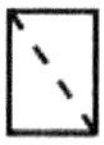	Dust or particle trap
	Funnel
	Steam trap
	Viewing glass
	Pressure reducing valve
	Flexible pipe
	Valve
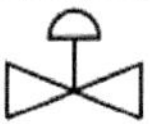	Control valve
	Manual valve
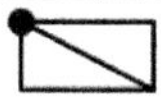	Back draft damper
	Needle valve
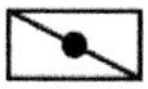	Butterfly valve
	Diaphragm valve
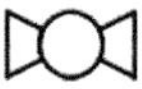	Ball valve
	Spring safety valve

Table 4.2 Connections types for P&IDs

Symbol	Name
	Pipe (process flow)
	Pneumatic signal
	Electrical signal
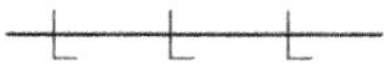	Hydraulic signal
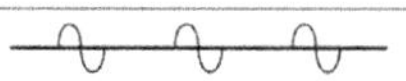	Electromagnetic Signal

Table 4.3 Location symbols

Location	Local/in the field	In a central location	In a local, central point
Symbol	PI 123.4	PI 123.4	PI 123.4
Comments	Component is found in the neighbourhood of the process itself	Component is found in some central location. Often, this central location is a computer	Component is found in some local central point, for example, a process-control cabinet

Table 4.4 Type symbols

	General	Process-Control Function
Symbol	PI 123.4	UY 123

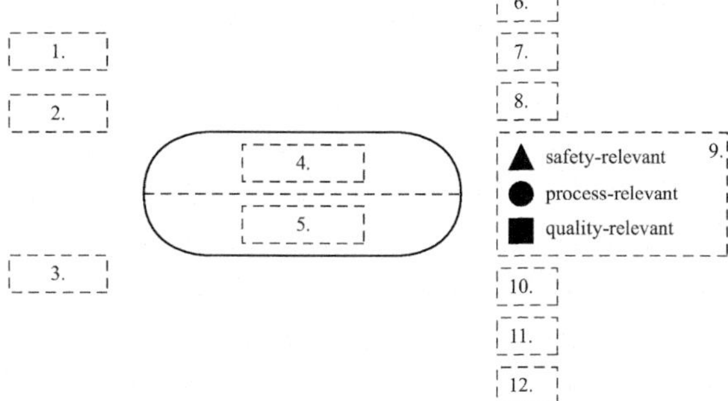

Fig. 4.7 Fields in a P&ID label

4.3.2 *Connections and Piping in P&IDs*

The type of connection must be shown in a P&ID. Table 4.2 shows the most common possibilities for such connections.

4.3.3 *Labels in P&IDs*

Another important feature of P&IDs is that the various components must be clearly identified. Sensors, valves, and other actuators should be named according to different guidelines. As well, the type and location of the components must be clearly shown. Tables 4.3 and 4.4 show the symbols that are used to show the location of actuators and sensors.

There are many different fields in the equipment label. These are shown in Fig. 4.7. The left fields (#1 to 3) are only used if needed, and there are no restrictions on what can be placed there. Field #1 often gives the supplier, while Field

Table 4.5 PCE categories

Letter	Meaning
A	Analysis
B	Burner combustion
C[3]	*Conductivity*
D	Density
E	Voltage
F	Flow
G	Gap
H	Hand
I	Current
J	Power
K	Time schedule
L	Level
M	Moisture
N	Motor
O	*Free*
P	Pressure
Q	Quantity/event
R	Radiation
S	Speed, frequency
T	Temperature
U	*Anticipated for PCE control functions*
V	Vibration
W	Weight
X	*Free*
Y	Valve
Z	*Free*

#2 gives the standard value of the component. The two central fields (#4 and 5) show the important information about the component. Field #4 shows the PCE category[4] and the PCE processing category (see Table 4.5 for additional information). Field #5 shows the PCE tag that is arbitrary. The right fields (#6 to 12) show additional information about the component. For example, Fields #6 to #8 show alarms and notifications related to upper limits, while Fields #10 to 12 contain alarms and notifications related to lower limits. The fields located furthest from the centre contain the most important information. Field #9 shows the importance of the component. A triangle (▲) shows a safety-relevant component, while a circle (●) shows a component required for the good manufacturing process (GMP). A square (■) shows a quality-relevant component.

Table 4.5 shows the PCE categories, while Table 4.6 shows the PCE processing functions. For motorised drives (PCE category N), there are only two possibilities

[3] Officially, this is a free variable to be set as needed. Practically, it is often used for conductivity.

[4] PCE is an abbreviation for *process-control engineering*.

for the processing function: S, which implies an on/off motor, and C, which implies motor control. For valves (PCE category Y), there are also limitations: S implies an on/off valve, C implies a control valve, Z implies an on/off valve with safety relevance, and IC implies a control valve with a continuous position indicator. For process-control functions, there are some very specific rules:

(1) The PCE category is always the same U.
(2) The following letters are often A, C, D, F, Q, S, Y, or Z. Combinations of these letters are also possible.

Example 4.2: P&ID Tags
What is the meaning of the following P&ID tags: PI-512, UZ-512 und MDI-512?

Solution

For PI-512, the first letter is "P". From Table 4.5, we can see that "P" represents pressure. The following letter is "I". From Table 4.6, we can see that "I" represents indicator. Thus, "PI-512" represents a pressure indicator.

For UZ-512, the first letter is "U". From Table 4.5, we can see that "U" represents a process-control function. The following letter is "Z". From Table 4.6, we can see that "Z" represents a safety-critical control function. Thus, "UZ-512" represents a safety-critical control function that is computed using a computer/PLC.

(continued on the next page)

Table 4.6 PCE processing categories

Letter	Meaning	Comment
A	Alarming	Only in Fields #6,7, 8, 10, 11, and 12; in Field #4 for process-control functions
B	Condition, limitation	Only in Field 4
C	Control	Only in Field 4
D	Difference	Only in Field 4
F	Fraction	Only in Field 4
H	High, on, open	Only in Fields #6, 7, 8, 10, 11, and 12
I	Indicator	Only in Field 4
L	Low, off, closed	Only in Fields #6, 7, 8, 10, 11, and 12
O	Local or PCS status indicator from a binary signal	Only in Fields #6, 7, 8, 10, 11, and 12
Q	Quantity	Only in Field 4
R	Recording	Only in Field 4
S	Switching	Only in Fields #6, 7, 8, 10, 11, and 12; in Field #4 for process-control functions
T	Transmitter	Only in Field 4
Y	Computation	Only in Field 4
Z	Emergency	Only in Fields #6, 7, 8, 10, 11, and 12; in Field #4 for process-control functions

For MDI-512, the first letter is "M". From Table 4.5, we can see that "M" represents moisture. The following letters are "DI". From Table 4.6, we can see that "D" represents difference and "I" represents indicator. Thus, "MDI-512" represents a moisture difference indicator.

The labels for the other fields are not standardised. However, in a P&ID, the same label should be used for the same component/idea. All P&IDs require a legend, where all the components are briefly described, including such information as their names, engineering data (size, material), and operating conditions. Appropriate stamps and signatures may need to be affixed depending on local laws, for example, in Canada, all engineering documents require a stamp and signature from a practising engineer.

4.4 Electric and Logic Circuit Diagrams

Since much automation is implemented using electrical signals and logic statements, it is useful to understand the basics of how such circuit diagrams are constructed. As well, some of these symbols will re-appear when considering graphical programming languages.

There exist two main standards for drawing the shapes in electric and logic circuit diagrams: the official DIN EN 60617 (a.k.a. the IEC 617) standard and the nonpreferred, but commonly encountered, ANSI IEEE 91-1991 standard. The official standard, although not explicitly showing the nonpreferred symbols, states that it is permissive to use locally accepted national versions of the symbols. Common symbols for both standards are shown in Table 4.7.

Furthermore, when it comes to showing connections, for example, the splitting of a wire, different formats are possible. The recommended approach is to use T-junctions to show a split (as shown by Fig. 4.8a). However, often, the split will be shown as in Fig. 4.8b. A simple crossing of wires is shown by Fig. 4.8c.

4.5 Further Reading

The following are additional resources on the topics mentioned in this chapter:

(1) **P&ID**: M. Toghraei (2019). *Piping and Instrumentation Diagram Development*, Hoboken, New Jersey, USA: Wiley.

Table 4.7 Common symbols in circuit diagrams

Item	DIN-60617 symbol	ANSI symbol	Comments
Resistor			
Inductor			
Direct-current voltage source		+	Essentially a battery
Alternating-current voltage source			
Diode			
Capacitor			
AND Gate	&		
OR Gate	≥1		
Inverter	1		
NAND Gate	&		
NOR Gate	≥1		
XOR Gate	=1		
Ground			
Variable			The arrow is placed over the element that is variable, for example, a variable resistor: &

(continued)

Table 4.7 (continued)

Item	DIN-60617 symbol	ANSI symbol	Comments
General negation	O	O	Negation is placed where the signal enters or leaves a block (see, for example, the NAND or NOR gates above)

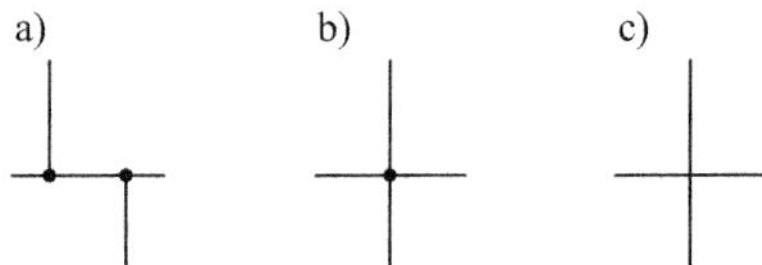

Fig. 4.8 Connections: a: recommended form for contact; b: commonly encountered form for contact; and c: no contact

4.6 Chapter Problems

Problems at the end of the chapter consist of three different types: (a) Basic Concepts (True/False), which seek to test the reader's comprehension of the key concepts in the chapter; (b) Short Exercises, which seek to test the reader's ability to compute the required parameters for a simple data set using simple or no technological aids; and (c) Computational Exercises, which require not only a solid comprehension of the basic material, but also the use of appropriate software.

4.6.1 Basic Concepts

Determine if the following statements are true or false and state why this is the case.

(1) A block diagram is a frequency-domain representation of the process.
(2) A pneumatic signal is denoted using a line with sinusoids.
(3) An electric signal is denoted using a dashed line.
(4) A tag placed in a round symbol with a single line through the middle represents a component located in some central location.
(5) An instrument with the tag TIC is a density-indicator controller.
(6) An instrument with the tag PIC is a pressure-indicator controller.
(7) An instrument with the tag RI is a radiation indicator.
(8) An instrument with the tag TDI is a temperature-difference indicator.
(9) An instrument with the tag JI is a current indicator.
(10) A P&ID should contain a legend explaining the symbols and notation used.

4.6.2 Short Questions

These questions should be solved using pen and paper. Appropriate software for drawing the required diagrams can also be used to assist with the design.

(11) Write the process model between U and Y for the block diagrams shown in Fig. 4.9.

(12) Draw the process-flow diagrams for the following process descriptions. *Hint: Knowledge of chemical engineering and the physical behaviour of systems is required.*

a. The DEA solution enters the treatment unit via a throttle valve and is heated in three consecutive heat exchangers. The heated solution enters the first evaporation column which is operated under vacuum at 80 kPa and a temperature of 150°C. Most of the water is evaporated here. The water vapour is used in the first heat exchanger to heat the incoming solution. The concentrated DEA solution is transferred from the first evaporator to a second evaporation column via a second throttle valve, so that the pressure can be lowered to 10 kPa. The second evaporation column is a thin-film evaporator (similarly built to a scraped-wall heat exchanger), which is heated by high-pressure steam. Here, most of the DEA is evaporated. The remaining slurry is removed from the column by a positive displacement pump. The DEA vapour is used to heat the feed in the second heat exchanger. The third heat exchanger in the feed line is heated by medium-pressure steam to achieve sufficient evaporation in the first evaporation column. After exchanging heat with the feed stream, the water vapour and DEA vapour streams are completely liquefied in two condensers operated with cooling water, before being pumped to a T-junction and returned to the gas scrubbing process.

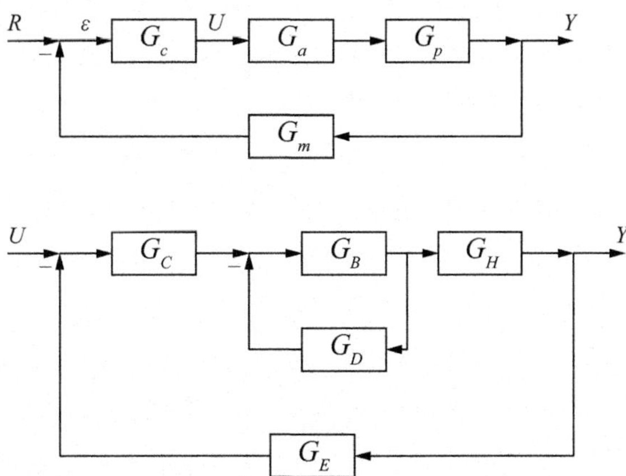

Fig. 4.9 Block diagrams for Question 11

b. A concentrated solution of 99% triethylene glycol ($C_6H_{14}O_4$) and 1% water is commonly used to absorb moisture from natural gas at 4.1 MPa. Natural gas is mainly methane, with smaller amounts of ethane, propane, carbon dioxide, and nitrogen. The water-saturated natural gas is contacted with the triethylene glycol at 40°C in a trayed absorber column to remove the water. The dried gas is then ready for piping to market. The triethylene-glycol solution leaving the absorber column is diluted by water, and must be regenerated for re-use. It first passes through a flash drum at a pressure of 110 kPa to release dissolved gases. Then, it is heated in a heat exchanger with the hot regenerated glycol. The dilute solution then enters the top of a packed regeneration column. As it flows downward over the packing, it passes steam flowing upwards which evaporates water from the glycol solution. The reboiler at the bottom of the regeneration column uses medium-pressure steam to heat the triethylene-glycol solution to 200°C to generate the vapour that passes up the column. At the top of the column, the vapour (*i.e.* the water evaporated from the triethylene-glycol solution) is vented to the atmosphere. The hot regenerated solution from the bottom of the regenerator column is cooled in a heat exchanger with the dilute solution from the absorber. It is then pumped through an air cooler and then back to the absorber column at approximately 45°C.

(13) Using the sketch of the process-flow diagram in Fig. 4.10 and the process description below, provide a properly formatted process-flow diagram that includes all the required components and information. The process description is as follows. Ethyl chloride (C_2H_5Cl) is produced in a continuous, stirred tank reactor (CSTR) filled with a slurry of catalyst suspended in liquid ethyl chloride. Most of the heat of reaction is absorbed by vaporising 25 kmol/hr of the liquid ethyl chloride. This vapour leaves the reactor with the product stream. The reactor is jacketed to allow additional temperature control if required. All the ethyl chloride in the product stream is condensed, and enough ethyl chloride is returned to the reactor to maintain steady state. The waste gas is sent to the flare system.

(14) Evaluate the P&IDs shown in Fig. 4.11. Have they been corrected drawn? Is there anything that is missing? Are all the streams correctly shown?

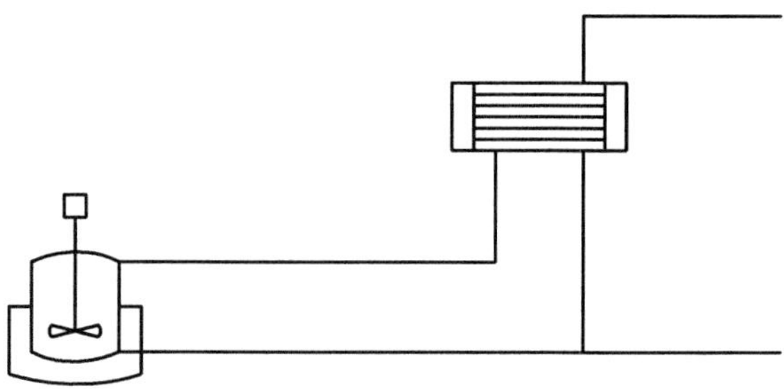

Fig. 4.10 Sketch of the process-flow diagram for Question 13

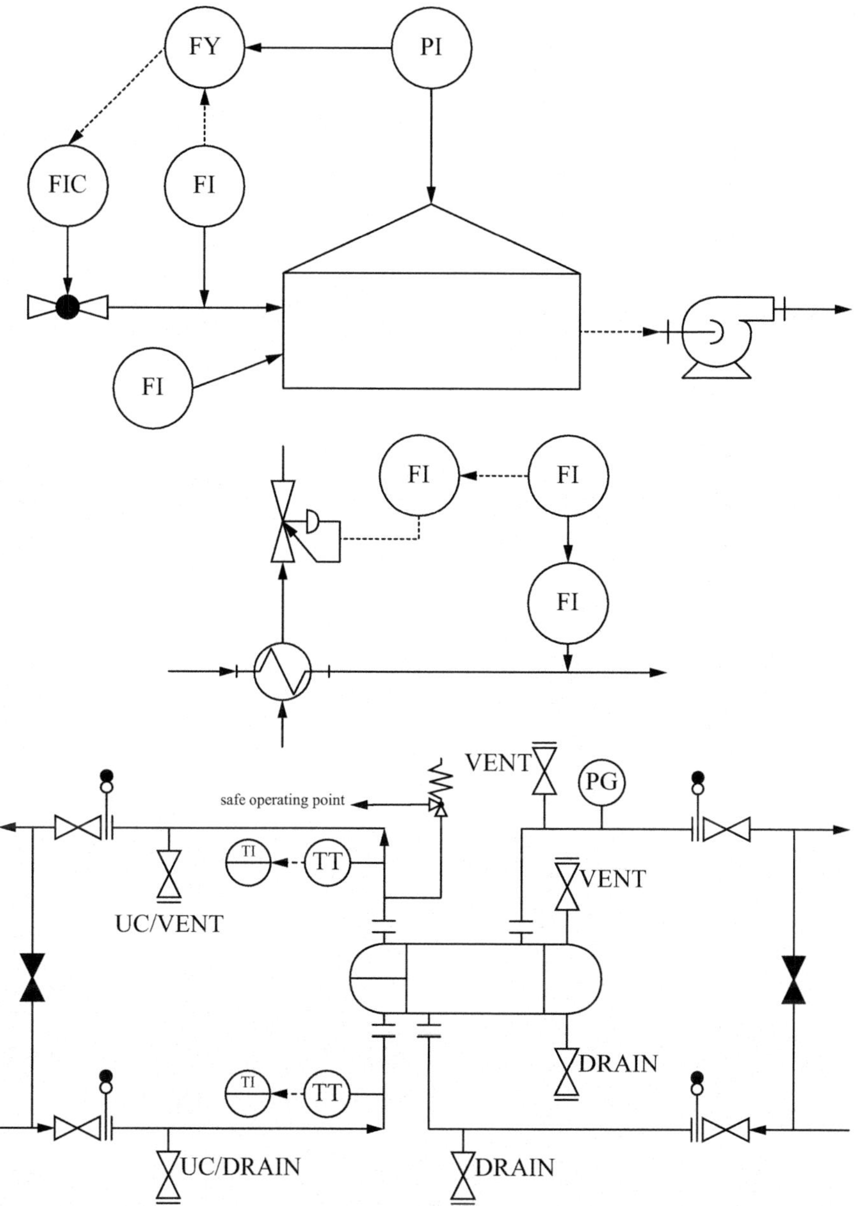

Fig. 4.11 P&IDs for Question 14

(15) Given the P&ID in Fig. 4.12, determine where the controller is located and what signals it requires.

(16) For the process shown in Fig. 4.13, answer the following questions:

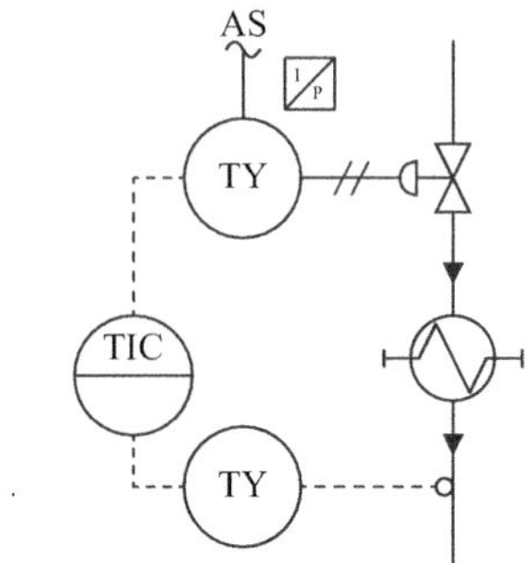

Fig. 4.12 P&ID for Question 15

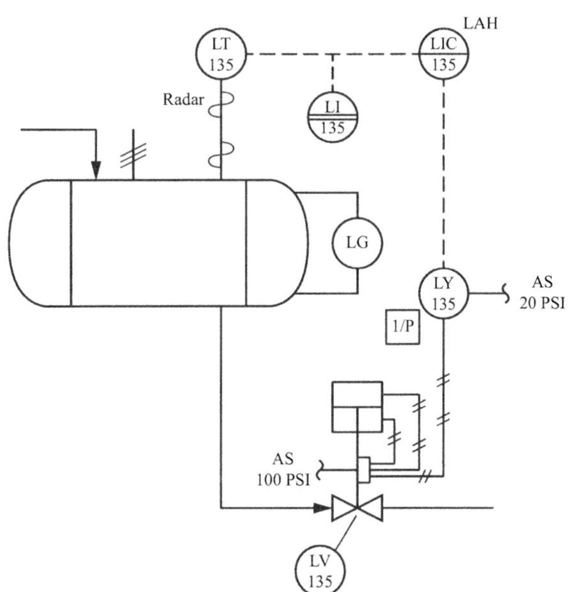

Fig. 4.13 P&ID for Question 16

a. What kind of alarm does the controller provide?
b. How is the level measured? What kind of signal is produced?
c. What kind of valve is used to control the level?
d. What do the double lines for item LI-135 mean?

(17) Imagine that you producing maple syrup. The P&ID along with the current values is shown in Fig. 4.14. Determine if there are any measurement errors.

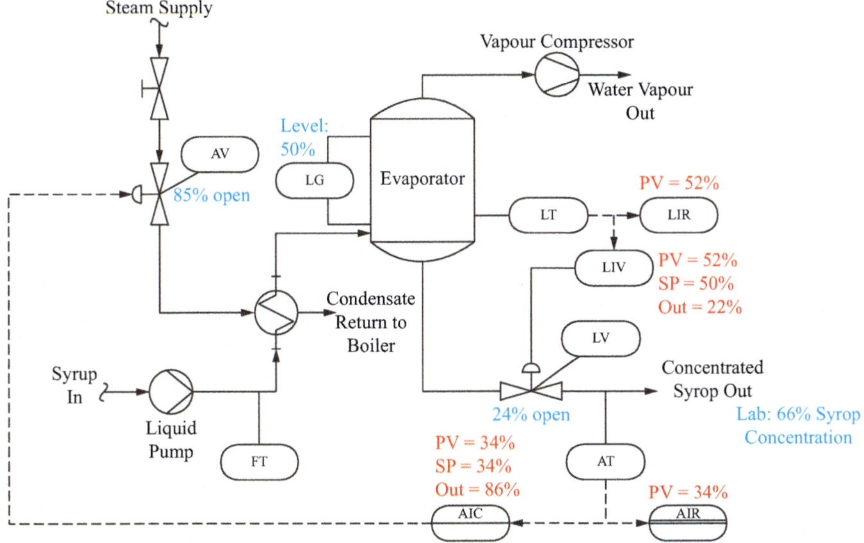

Fig. 4.14 P&ID for the production of maple syrup for Question 17

4.6.3 Computational Exercises

The following problems should be solved with the help of a computer and appropriate software packages.

(18) Take a complex process and draw the P&ID for it. Make sure to include all the relevant sensors and controllers.

(19) Using the P&ID from Question 18, create a simplified process-flow diagram for the process.

Chapter 5
Control and Automation Strategies

Having looked at the different instruments that can be used for automation and how to describe the process and its interconnections, it is now necessary to see how the desired automation can actually be implemented. The **control** or **automation strategy** is the method used to control/automate the process to achieve the desired objectives. Although two different words can be used here, the overall concepts are very similar: produce a system that can operate on its own with minimal human interference. Depending on the field and even industry, one of these two words may be more common.

There exist many different ways to classify the different types of strategies that can be implemented. Often, in practice, these strategies can be combined to produce a final overall control strategy. The following are the most common types of control strategies:

(1) **Open-Loop Control** (or steering, servoresponse): In open-loop control, the desired trajectory is set and implemented. The object follows this trajectory without taking into consideration any variations in the environment or surroundings. Clearly, if there are any changes in the environment, the object may not achieve its trajectory.

(2) **Closed-Loop Control** (or regulation, regulatory response): In closed-loop control, the actual value of the system is always compared against the setpoint value. Any deviations are then corrected using a controller. This allows for the system to make corrections based on imperfections in the original specifications, changes in the environment, or unexpected events. In most control strategies, some aspect of closed-loop control will be implemented.

(3) **Feedforward Control**: In feedforward control, known disturbances are measured and corrective action is taken so that the disturbance does not affect the plant. This is a way of taking into consideration future information about the process and taking action now.

Y. A. W. Shardt, *Automation Engineering*,
https://doi.org/10.1007/978-3-031-92394-4_5

(4) **Discrete-Event Control**: In discrete-event control, control only occurs, when some logic condition is triggered. This is often used in safety-relevant systems to trigger immediate action or raise an alarm, for example, if the pressure is above a certain threshold, then the pressure release valve should be opened. Often, such systems are modelled using automata.

(5) **Supervisory Control**: In supervisory control, the control loop does not control a process, but it controls another control loop. Supervisory control loops are common in complex industrial systems, where there may be multiple objectives and variables to control.

5.1 Open- and Closed-Loop Control

Since the two control strategies are often combined together, it is useful to understand the difference between them.

5.1.1 *Open-Loop Control*

Consider a system as shown in Fig. 5.1, where the manipulated variable $u(t)$ influences the output variable, $y(t)$. The desired behaviour is specified by the setpoint $r(t)$. The path from $r(t)$ to $u(t)$ to $y(t)$ is called the open-loop control path. The objective here is to design a controller, G_c, such that given a setpoint trajectory, it can produce a sequence of value for the manipulated variable $u(t)$, so that the output $y(t)$ attains the values specified by $r(t)$. The disturbance affecting the system is given as $e(t)$. Often, it will be assumed that $e(t)$ is a Gaussian, white-noise signal. The transfer function for the disturbance is G_d.

In principle, if we are given a mathematical function that converts $u(t)$ into $y(t)$, more commonly called a plant model and denoted by G_p, then using the inverse of this model as the controller will allow us to attain the goal of making $y(t) = r(t)$. However, in practice, the following issues arise:

1. **Unreliability of the Model Inverse**: Due to time delays and unstable zeros of the process, it may not be possible to obtain a realisable inverse model.
2. **Plant-Model Mismatch**: It is very likely that the model for G_p is not exact, which implies that the incorrect parameters are given to the controller leading to incorrect results.

Fig. 5.1 Open-loop control

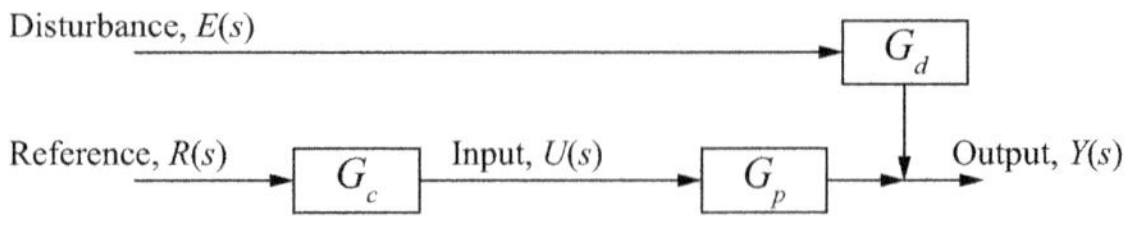

Disturbance, $E(s)$
G_d
Reference, $R(s)$
G_c
Input, $U(s)$
G_p
Output, $Y(s)$

3. **Disturbances**: In reality, there will be disturbances that prevent $y(t)$ from reaching the value specified by $r(t)$. One solution to this problem is to implement feedforward control that measures the disturbance and takes corrective action. However, this only solves the problem if the disturbance variable can be measured.

Example 5.1: Temperature Control: Open-Loop Control

Figure 5.2 shows a diagram for the open-loop control of the temperature in the house, where the radiator setting is used to determine the temperature in the room. However, once the radiator setting is set, no further actions are taken if a disturbance should occur, for example, the temperature outside changes, more people enter the room, the window is opened, or the sun starts shining. All of these disturbances will change the temperature in the room, but no automatic change will occur in radiator setting. Of the disturbances, the largest is probably the outside temperature. If, as shown in Fig. 5.2, we can measure the outside temperature, then we can design a feedforward controller to correct for the disturbance caused by this variable.

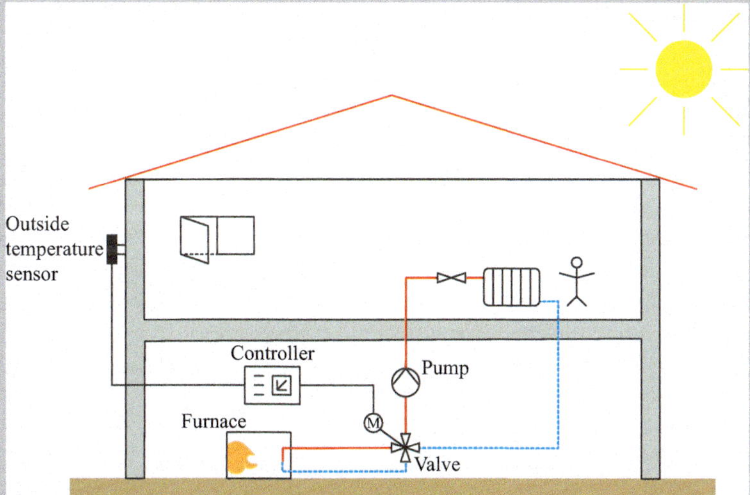

Fig. 5.2 Open-loop control for the temperature in the house

5.1.2 Closed-Loop Control

In closed-loop control, information about the process is used to update the control action. Figure 5.3 shows the overall closed-loop control system, where the difference between the reference signal and the actual output is used as the input into the controller. The controller error, $\varepsilon(t)$, is defined as the difference between the setpoint and the measured output. This error is then the input into the controller. In order to emphasise that the output must be measurable in order for closed-loop control to occur, the sensor transfer function, G_m, is explicitly shown. Similarly, to emphasise the importance of the actuator, its transfer function is also explicitly included as G_a. In most applications, it will be assumed that G_m and G_a have a minimal effect on the overall system, and hence, they will be ignored.

Closed-loop control provides the following advantages over open-loop control:

(1) **Better accuracy**, especially since it can compensate for unknown and unmeasurable disturbances or imprecise models of the system.
(2) **Stability**, since closed-loop control can stabilise unstable processes.

On the other hand, closed-loop control requires the ability to measure the output variables, which can lead to a greater cost. As well, a poorly designed closed-loop controller can destabilise the system.

Example 5.2: Temperature Control: Part 2: Closed-Loop Control
Consider the same set-up as in Example 5.1, but modified as shown in Fig. 5.4, so that the temperature in the room can now be measured and an appropriate controller designed. Based on the temperature in the room, the controller can set the valve opening which will determine the flow of heat into the room. Opening the valve more will cause more heat to enter the room, thus increasing its temperature. Closing the valve will cause less heat to enter the room, thus decreasing its temperature.

(continued on the next page)

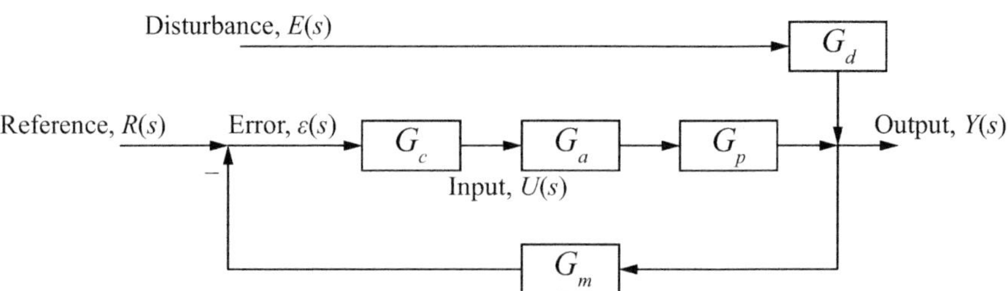

Fig. 5.3 Closed-loop control

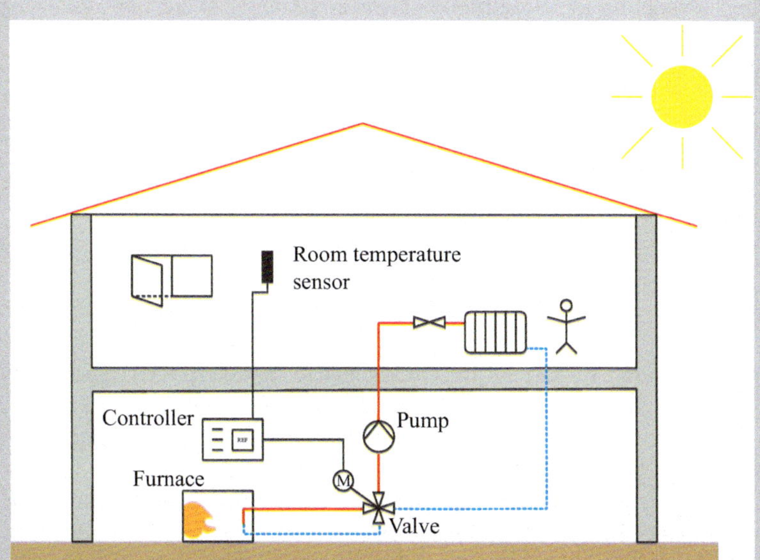

Fig. 5.4 Closed-loop control of the temperature in the house

Furthermore, other disturbances may occur that are not easily measured, but whose effect can be controlled by the closed-loop controller. Figure 5.4 shows people in the room, an open window, and direct sunlight. Since these will influence the room temperature, their effect will be noticed by the room-temperature sensor that will cause the controller to make appropriate changes to maintain the temperature at its desired value.

In closed-loop control, the following are the key ideas to consider (Seborg, Edgar, Mellichamp, & Doyle, 2011):

(1) The closed-loop system should be **stable**; otherwise, the system is worse than before.
(2) The effect of **disturbances** should be minimised.
(3) The controller should provide quick and effective change between different setpoints, that is, **setpoint tracking** should be good.
(4) There should not be any **bias**, that is, the output should reach the desired value.
(5) Large **actuator values** should be minimised.
(6) The controller should be **robust**, that is, small changes in the process characteristics should not make the overall system unstable.

When designing a closed-loop control strategy, there are two common approaches to take:

(1) **State Feedback**, where it is assumed that the controller is a constant K and all the states are measurable and available, that is, $y_t = x_t$. The control law is given as $u_t = -Kx_t$.

(2) **Proportional, Integral, and Derivative (PID) Control**, where the control law can be written as

$$U(s) = K_c\left(1 + \frac{1}{\tau_I s} + \tau_D s\right)\varepsilon(s) \tag{5.1}$$

where K_c, τ_I, and τ_D are parameters to be determined. There exist different forms of the PID control law depending on the industry, era, and implementation. It is common to drop the D-term, that is, set $\tau_D = 0$, to give a PI controller.

In addition to these common methods, there also exist various nonlinear approaches that can be applied to improve overall control, such as **deadbanding** or **squared-error control**.

State Feedback

A state-feedback controller is an effective manner for designing a controller when it is possible to determine the states. A state-feedback controller consists of two parts: an **observer** and a **controller**. The observer uses the output from the process to estimate the states, while the controller uses the (estimated) states to determine the input. Figure 5.5 shows a block diagram of such a state-feedback controller and observer.

It is possible to design different types of observers, depending on the intended application. The most common observer is the **Luenberger observer** which seeks to minimise the error between the predicted and actual states. Assume that the Luenberger observer can be written as

$$\begin{aligned}\dot{\hat{\vec{x}}} &= \mathcal{A}\hat{\vec{x}} + \mathcal{L}\left(\vec{y} - \hat{\vec{y}}\right) + \mathcal{B}\vec{u} \\ \hat{\vec{y}} &= \mathcal{C}\hat{\vec{x}} + \mathcal{D}\vec{u}\end{aligned} \tag{5.2}$$

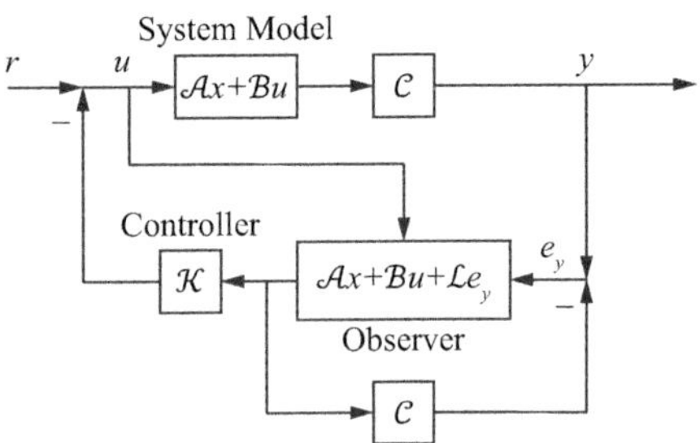

Fig. 5.5 State-feedback control

where $\mathcal{L}$, the **observer gain**, is an appropriately sized matrix that needs to be designed and $\hat{\circ}$ represents an estimated value. Rewriting Eq. (5.2) gives

$$\begin{aligned}\dot{\hat{\vec{x}}} &= (\mathcal{A} - \mathcal{L}\mathcal{C})\hat{\vec{x}} + \mathcal{L}\vec{y} + (\mathcal{B} - \mathcal{L}\mathcal{D})\vec{u} \\ \hat{\vec{y}} &= \mathcal{C}\hat{\vec{x}} + \mathcal{D}\vec{u}.\end{aligned} \tag{5.3}$$

Let the error be defined as

$$\vec{e} = \vec{x} - \hat{\vec{x}}. \tag{5.4}$$

Taking the derivative (or difference) of Eq. (5.4) gives

$$\dot{\vec{e}} = \dot{\vec{x}} - \dot{\hat{\vec{x}}}. \tag{5.5}$$

Substituting Eqs. (3.30) and (5.3) into Eq. (5.5) gives

$$\begin{aligned}\dot{\vec{e}} &= \mathcal{A}\vec{x} + \mathcal{B}\vec{u} - (\mathcal{A} - \mathcal{L}\mathcal{C})\hat{\vec{x}} - \mathcal{L}(\mathcal{C}\vec{x} + \mathcal{D}\vec{u}) - (\mathcal{B} - \mathcal{L}\mathcal{D})\vec{u} \\ &= (\mathcal{A} - \mathcal{L}\mathcal{C})\vec{e}.\end{aligned} \tag{5.6}$$

From Eq. (5.6), it can be seen that for the error to reach zero $\mathcal{A} - \mathcal{L}\mathcal{C}$ must be stable, that is, the eigenvalues of $\mathcal{A} - \mathcal{L}\mathcal{C}$ must be located so as to make the process stable.[1]

Now that we can estimate the state values, it makes sense to design a **state controller**. Assume that the state control law can be written as[2]

$$\vec{u} = -\mathcal{K}\hat{\vec{x}}. \tag{5.7}$$

Before examining the general case, let us consider the situation, where perfect information can be obtained about the states and there is no need to design an observer, that is, $\mathcal{C} = \mathcal{I}$, which implies that $\vec{y} = \vec{x}$. In this situation, we can use the basic state-space equations, given by Eq. (3.30), to describe the system. Substituting Eq. (5.7) into Eq. (3.30) and re-arranging gives

$$\begin{aligned}\dot{\vec{x}} &= \mathcal{A}\vec{x} - \mathcal{B}(\mathcal{K}\vec{x}) = (\mathcal{A} - \mathcal{B}\mathcal{K})\vec{x} \\ \vec{y} &= \mathcal{C}\vec{x} + \mathcal{D}\vec{u}.\end{aligned} \tag{5.8}$$

Thus, it can be seen that the stability of the controlled (overall, closed-loop) system is determined by the location of the eigenvalues of the $\mathcal{A} - \mathcal{B}\mathcal{K}$ matrix.

Now consider the case, where we do not have full information about the process. In this case, we will need to consider two different equations: the observer equation and the true state-space model. Substituting the control law given by Eq. (5.7) into the observer equation, Eq. (5.3), gives

---

[1] Continuous domain: $\mathrm{Re}(\lambda) < 0$; discrete domain: $\|\lambda\| < 1$.

[2] Often, the estimated state is replaced by the true state value. As it will be shown, this makes sense in many cases since we have full information about the states by actually measuring them.

$$\begin{aligned}\dot{\hat{\vec{x}}} &= (\mathcal{A} - \mathcal{L}\mathcal{C} - \mathcal{B}\mathcal{K} + \mathcal{L}\mathcal{D}\mathcal{K})\hat{\vec{x}} + \mathcal{L}\vec{y} \\ \hat{\vec{y}} &= \mathcal{C}\hat{\vec{x}} - \mathcal{D}\mathcal{K}\hat{\vec{x}}.\end{aligned} \tag{5.9}$$

The fact that both the $\mathcal{L}$- and $\mathcal{K}$-matrices are present in this equation raises the question if the controller and observer must be designed simultaneously or could they be designed separately.

In fact, the linear, state-space law presented here coupled with the linear observer are fully separable and stability will be assured if both components are stable. In order to show this, rewrite the controller law as

$$\vec{u} = -\mathcal{K}(\vec{x} - \vec{e}). \tag{5.10}$$

From Eq. (5.8), we can write the true state-space system as

$$\dot{\vec{x}} = \mathcal{A}\vec{x} - \mathcal{B}\mathcal{K}(\vec{x} - \vec{e}) = (\mathcal{A} - \mathcal{B}\mathcal{K})\vec{x} + \mathcal{B}\mathcal{K}\vec{e}. \tag{5.11}$$

Now, combining this equation with the error equation, given as Eq. (5.6), into a single equation, we can see that the combined system consisting of the states and errors can be written as

$$\begin{bmatrix}\dot{\vec{x}} \\ \dot{\vec{e}}\end{bmatrix} = \begin{bmatrix}\mathcal{A} - \mathcal{B}\mathcal{K} & \mathcal{B}\mathcal{K} \\ 0 & \mathcal{A} - \mathcal{L}\mathcal{K}\end{bmatrix}\begin{bmatrix}\vec{x} \\ \vec{e}\end{bmatrix}. \tag{5.12}$$

Since the combined system matrix is triangular, the eigenvalues can be obtained by considering the eigenvalues of the diagonal elements, that is, the eigenvalues of the $\mathcal{A} - \mathcal{B}\mathcal{K}$ and $\mathcal{A} - \mathcal{L}\mathcal{K}$ matrices. Furthermore, since neither of the two matrices interact with each other, it can be concluded that both can be designed separately and then combined. As long as each is stable, then the overall system will also be. This result is called the **separation principle** and is a very powerful result in designing state-feedback systems.

Proportional, Integral, and Derivative (PID) Control

Another class of common controllers is the proportional, integral, and derivative (PID) controllers. This class of controllers provides effective control for many industrial processes and can be configured to deal with a wide range of different situations. The general form of the PID controller can be written as

$$G_c = K_c\left(1 + \frac{1}{\tau_I s} + \tau_D s\right) \tag{5.13}$$

where K_c is the **proportional term** or **controller (proportional) gain**, τ_I is the **integral term** or the **integral time constant** in units of time, and τ_D is the **derivative term** or the **derivative time constant** in units of time. Another common representation is given as

$$G_c = K_c + K_I\frac{1}{s} + K_D s \tag{5.14}$$

where K_I is the integral gain and K_D is the derivative gain. In many industrial plants, it is possible to find the terms **reset** for integral and **rate** for derivative, for example, τ_I would be called the reset time constant.

In industry, different combinations of this controller can be found including the rather common proportional and integral controller (PI), where $\tau_D = K_D = 0$, that is, the D-term is ignored. In order to understand the behaviour of this controller, each of the three terms (P, I, and D) will be investigated separately. As well, the two common configurations (PI and PID) will be considered in greater detail.

Proportional Term

The proportional term, K_c, has the greatest effect on the overall control performance. It represents the contribution of the current error to the overall controller action, and thus, controls two key aspects: stability and speed of response. The larger the absolute value of K_c is, the faster the system will response to changes, but the less stable the system will be. Furthermore, a controller containing only the proportional term will have bias or offset. Applying the final-value theorem to the closed-loop transfer function for the setpoint derived in Example 4.1 and ignoring the actuator and sensor transfer functions gives

$$\lim_{s\to 0}\frac{sG_cG_p}{1+G_cG_p}R_s = \lim_{s\to 0}\frac{sK_cG_p}{1+K_cG_p}\frac{M_r}{s} = \frac{K_cK_pM_r}{1+K_cK_p} \tag{5.15}$$

where R_s is the setpoint which gives a step change of M_r and K_p is the process gain. It is assumed that the overall closed-loop transfer function is stable. From Eq. (5.15), it can be seen that the steady-state value for the closed-loop transfer function is not M_r. This implies that the final value will not be equal to the setpoint. For this reason, pure P-controllers are rarely used.

Example 5.3: Investigation of the Proportional Term on Stability and Performance

Consider the following first-order system

$$G_p = \frac{1.5}{20s+1}e^{-10s} \tag{5.16}$$

controlled with a proportional controller $G_c = K_c$. Setting K_c equal to -0.25, 0, 0.5, 1.0, and 2.0, show how the closed-loop response to a step change of two in the setpoint changes.

Solution

Figure 5.6 shows the results for changing the K_c. It can be seen that for none of the cases is the setpoint attained. There is offset for all of the cases. Furthermore, for $K_c = 0$, the process remains at zero, since no control is being implemented. As K_c increases, the amount of oscillation and length of time required to attain a steady-state value increase. In fact, if K_c is increased further, the process will become unstable. Similarly, for $K_c < 0$, the process is initially stable. However, increasing the value will rapidly make the system unstable.

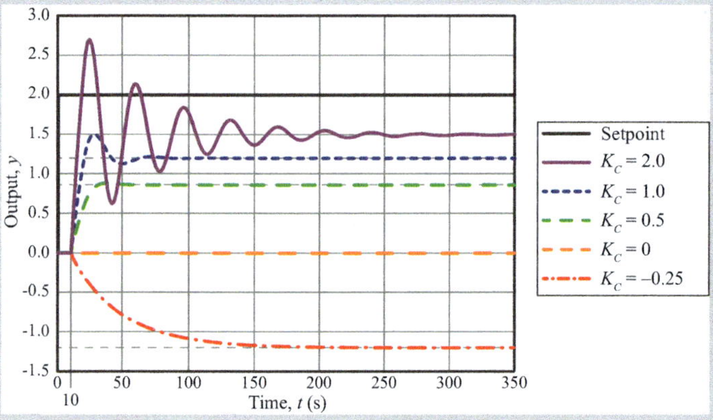

Fig. 5.6 Effect of changing K_c for a P-controller

Integral Term

The integral term, τ_I, provides a measure of additional stability to the system as well as removing any bias present. It represents the contribution of past errors on the current controller action. In general, the integral term is always combined with a proportional term to give the proportional and integral controller. Furthermore, a controller containing only the integral term will have no bias or offset, that is,

$$\lim_{s \to 0} \frac{sG_cG_p}{1 + G_cG_p} R_s = \lim_{s \to 0} \frac{s\frac{1}{\tau_I s}G_p}{1 + \frac{1}{\tau_I s}G_p} \frac{M_r}{s} = \lim_{s \to 0} \frac{M_rG_p}{\tau_I s + G_p} = M_r \tag{5.17}$$

where R_s is the setpoint which makes a step change of M_r. It is assumed that the overall closed-loop transfer function is stable. From Eq. (5.17), it can be seen that the closed-loop gain is exactly equal to the step change in the setpoint, which implies that there will be no bias or offset.

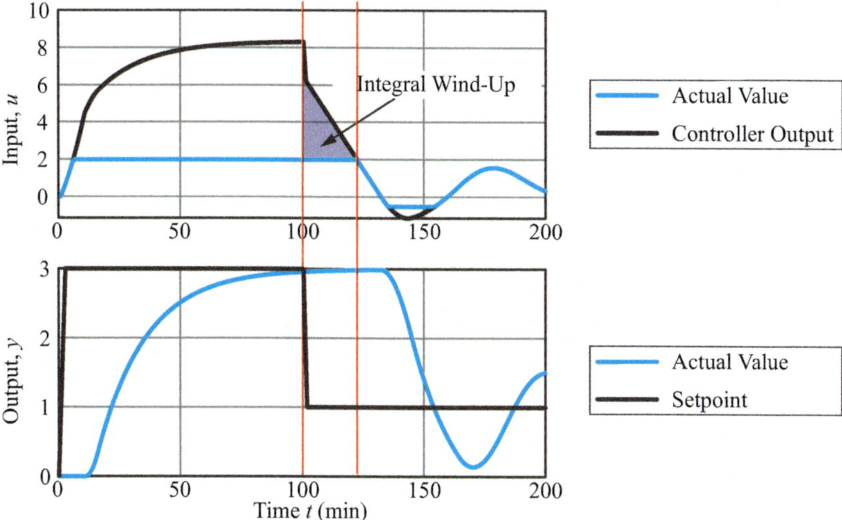

Fig. 5.7 Integral wind-up

Since the integral term represents the effect of past errors on the current controller action, it is possible that in a real implementation, **integral (reset) wind-up** can occur. This occurs when the controller output is above the physical limit of the actuator. Since the controller does not have a way of knowing that the physical limit has been reached, it continues to increase the desired controller output as the error keeps on increasing. This makes the integral term very large. When there is a need to now decrease this value, the controller decreases the output value from its large, unphysical value until it goes below the upper limit. The time delay between when the actuator should have started to respond and when it is observed to be responding is called integral wind-up. This is shown in Fig. 5.7, where it can be seen that at 100 min, when the setpoint is changed, it takes the process another 20 min before it even starts changing, since the true actuator value remains saturated at its upper limit. Integral wind-up can be avoided by including the physical limits of the actuator in the controller, that is, by making values beyond the physical limits of the actuator equal to the physical limit.

Example 5.4: Investigation of the Integral Term on Stability and Performance

Consider the following first-order system

$$G_p = \frac{1.5}{20s + 1} e^{-10s} \tag{5.18}$$

controlled with a proportional controller $G_c = 1/(\tau_I S)$. Changing the value of τ_I from 20 to 100 in increments of 10, show how the closed-loop response to a step change of two in the setpoint changes.

Solution

Figure 5.8 shows the results. It can be seen that as τ_I increases the effect on the system decreases. Note that the system is originally stable so that without any control it will reach the desired setpoint. Furthermore, and unlike with only a *P*-term, there is no bias in the closed-loop system.

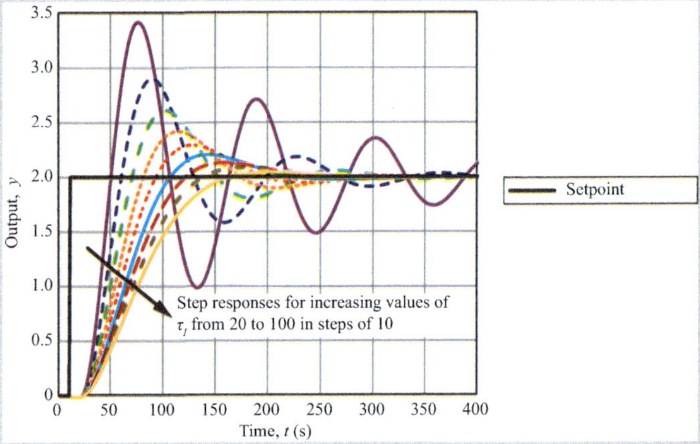

Fig. 5.8 Effect of changing τ_I for an I-controller

Derivative Term

The derivative term, τ_D, provides a measure of additional stability to the system. The derivative term represents the contribution of future (estimated) errors on the controller action. In general, the derivative term is almost always combined with proportional and integral terms. Furthermore, a controller containing only the derivative term will be unable to track the setpoint, that is,

$$\lim_{s\to 0}\frac{sG_cG_p}{1+G_cG_p}R_s = \lim_{s\to 0}\frac{s\tau_D sG_p}{1+\tau_D sG_p}\frac{M_r}{s} = \lim_{s\to 0}\frac{sM_rG_p}{1+\tau_D sG_p} = 0 \tag{5.19}$$

where R_s is the setpoint which makes a step change of M_r. It is assumed that the overall closed-loop transfer function is stable. From Eq. (5.19), it can be seen that irrespective of the change in the setpoint, the system will return to zero, which implies that the derivative term cannot track the setpoint. Therefore, the main role of the derivative term is for disturbance rejection, since in that case, it is desired that the system return to the same initial value.

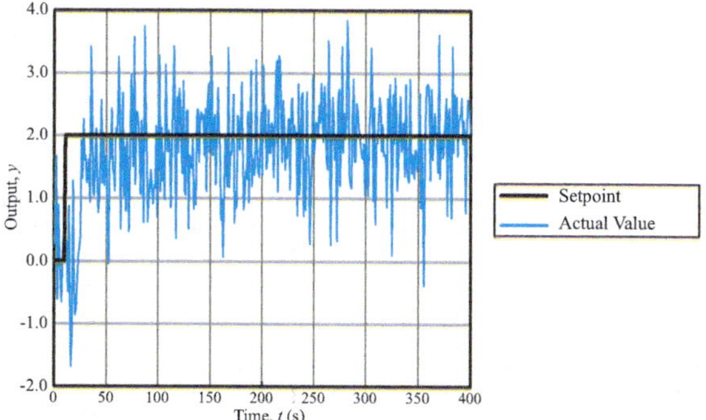

Fig. 5.9 Jitter with a derivative term

Furthermore, the derivative term as written cannot be physically realised, since the numerator is greater than the denominator. In order to avoid this problem, it is common to rewrite the derivative term into the following form:

$$G_{c,\text{real}} = \tau_D \frac{N}{\left(1 + \frac{N}{s}\right)} \tag{5.20}$$

where N is a filter parameter to be selected. It can be noted that as $N \rightarrow \infty$, this form will become equal to the original version.

Using a derivative term introduces two potential problems into the control strategy: **jitter** and **derivative kick**. Jitter occurs when the error signal fluctuates greatly about some mean value causing the estimated value of the future error to also fluctuate greatly. This in turn causes unnecessary fluctuations in the controller output, which can lead to issues with the actuator. Figure 5.9 shows the typical situation where jitter is present. Jitter can be mitigated by filtering the error signal used in computing the derivative term with a low-pass filter, which removes the high frequency component of the noise.

Derivative kick occurs when the setpoint has a sudden change in value (for example, a step change) causing the instantaneous error to be practically speaking infinite. This causes the controller error to spike for a very short period of time after the change occurs. Figure 5.10 shows the typical situation when derivative kick occurs. When the setpoint changes at 10 s, there is a corresponding spike in the actuator value. Note that in practice, due to saturation in the actuator, the effect of the derivative kick will be less than shown. Nevertheless, it will be present. Derivative kick can be lessened by solely using the change in the output rather than the error when computing the derivative term, that is,

$$u_t = -\tau_D s y_t. \tag{5.21}$$

Since it has been shown that the overall effect of the derivative term is primarily for disturbance rejection, that is, making $y_t \approx 0$, this approach is reasonable.

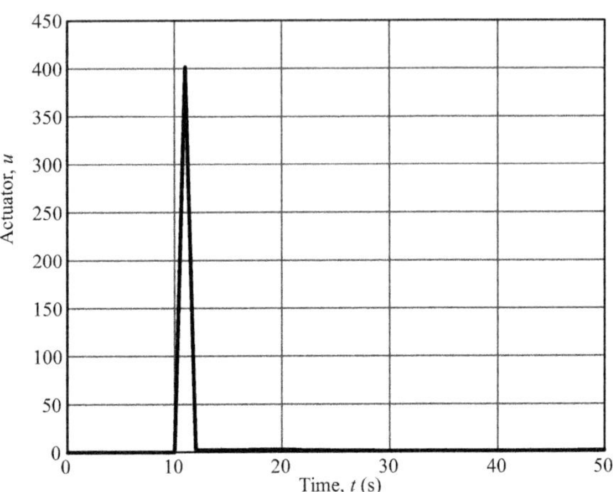

Fig. 5.10 Derivative kick

Example 5.5: Investigation of the Derivative Term on Stability and Performance

Consider the following first-order system

$$G_p = \frac{1.5}{20s+1}e^{-10s} \tag{5.22}$$

controlled with a derivative controller $G_c = \tau_D s$. Changing the value of τ_D from 1 to 11 in increments of 2 shows the closed-loop response of the controller to a white-noise disturbance of magnitude 0.05.

Solution

Figure 5.11 shows the results. It can be seen that as the magnitude of the *D*-Term increases, so does the magnitude of the response. This implies that the *D*-Term only influences the magnitude of the response.

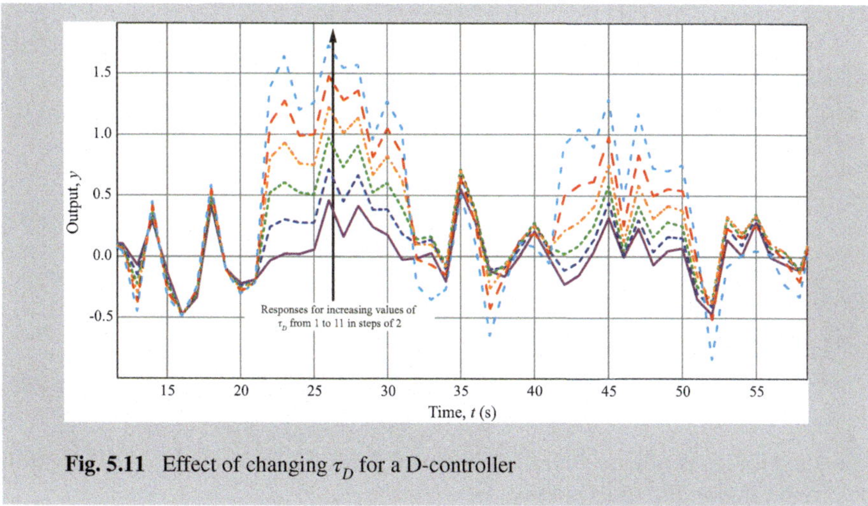

Fig. 5.11 Effect of changing τ_D for a D-controller

Proportional and Integral (PI) Controller

A proportional and integral (PI) controller is one of the most common industrial controllers found. It combines the proportional and integral terms to give

$$G_c = K_c\left(1 + \frac{1}{\tau_I s}\right). \tag{5.23}$$

The behaviour of a PI controller will be similar to that of each of the two separate terms, that is, the proportional term will influence the stability and overall performance, while the integral term will provide a biasfree controller. This controller can be interpreted as only considering the current and past information about the process and how that affects the overall control performance.

PI control can be used for both disturbance rejection and setpoint tracking in many different applications including flows, levels, and temperatures. Large time delays and highly nonlinear systems can limit its effectiveness.

Proportional, Integral, and Derivative (PID) Controller

A proportional, integral, and derivative (PID) controller can be tuned on the basis of the three previously considered terms. Such a controller takes all available information about the process and predicts the future trajectory of the system. It is commonly used in large complex systems where there is a need for an additional degree-of-freedom to stabilise the system. Systems which have a very jittery output or setpoint would need to be filtered before being used as the unfiltered signal can cause undesired behaviour.

Discretisation of the PID Controller

Since many control algorithms are implemented on a computer, it can often be helpful to implement the PID controller in the discrete domain. To discretise the PID controller, let $s = 1 - z^{-1}$ in Eq. (5.13) to give

$$u_k = K_c\left(1 + \frac{1}{\tau_I\left(1 - z^{-1}\right)} + \tau_D\left(1 - z^{-1}\right)\right)\varepsilon_k. \tag{5.24}$$

Re-arranging Eq. (5.24) gives

$$u_k = \frac{K_c}{\left(1 - z^{-1}\right)}\left(\left(1 - z^{-1}\right) + \frac{1}{\tau_I} + \tau_D\left(1 - z^{-1}\right)^2\right)\varepsilon_k. \tag{5.25}$$

Equation (5.25) is often called the **positional form** of the discretised PID controller. Re-arranging Eq. (5.25) to give

$$\left(1 - z^{-1}\right)u_k = \Delta u_k = K_c\left(\left(1 - z^{-1}\right) + \frac{1}{\tau_I} + \tau_D\left(1 - z^{-1}\right)^2\right)\varepsilon_k \tag{5.26}$$

produces another common PID equation called the **velocity form** of the discretised PID controller. In industrial implementations, it is possible to encounter both forms. Although these forms do not have an effect on controller design, they will have an effect on understanding the values obtained.

Controller Tuning

Having established the different types of controllers that can be used and understood some of behaviour of the parameters on the overall control system, the remaining question is how do we correlate a given set of performance criteria (for example, the objectives of control previously mentioned) and the different controller parameters. One easy approach would be to use some simulation software to obtain the desired parameter values using trial and error. As one can easily see, this approach could take very long and not necessarily provide the optimal parameter values. Instead, various correlations or approaches have been developed that will allow for the performance of a controller to be specified. In this section, we will consider the methods for tuning state-space controllers and PID controllers. Irrespective of the approach used, there will always be a need to simulate the resulting control system to make sure that there are no issues with the performance. Furthermore, it should be noted that for certain applications, there exist specific controller tuning rules that provide optimal performance for the problem at hand, for example, level controllers are often tuned using special rules that seek to minimise large deviations from the setpoint.

The general approach to controller tuning can be summarised by Fig. 5.12. Before starting the tuning procedure, it is required that the process be understood or modelled, that the performance criteria be clearly specified, and that the control strategies of interest be known. The first step will be to determine the initial controller parameters, followed by a simulation (or actual implementation, if the system can handle repeated perturbations). Based on the test and a comparison with the desired performance, new controller parameters can be obtained and new tests performed. This would be repeated until either the desired criteria are satisfied or the time available has run out. At this point, it may happen that the process understanding is found to be deficient and additional information would need to be sought. If the controller has not been implemented on the real system during the initial controller testing, then it will be necessary to repeat this procedure once more on the actual system. Very often, it will be found that the performance may not be as expected given the differences between the simulated and actual processes. However, following this procedure will minimise the risk that such a situation occurs.

Tuning a State-Space Controller

A state-space controller is often tuned using pole placement, which involves selecting the desired poles of the closed-loop transfer function and designing a controller that can achieve this.

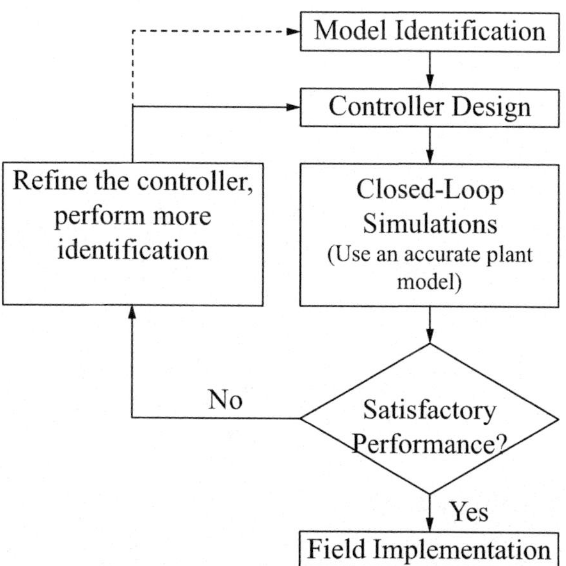

Fig. 5.12 Workflow for controller tuning

When placing the poles or eigenvalues of the closed-loop transfer function using a state-space approach, both the controller and observer will need to have their poles placed. The general rule of thumb is that the poles of the observer should be 10 to 20 times faster than the poles of the controller. Pole placement can be achieved using two different methods: **characteristic polynomial** and **Ackermann's Formula**. The approaches work for both continuous and discrete controllers. It should be noted that for multi-input systems, the computed value of $\mathcal{K}$ is not necessarily unique.

The general problem statement is: given the desired eigenvalues $\{\lambda_1, \lambda_2, \ldots, \lambda_n\}$ and the system $\{\mathcal{A}, \mathcal{B}, \mathcal{C}\}$, determine the state feedback controller gain $\mathcal{K}$. For the characteristic polynomial approach, perform the following steps:

(1) Compute $\overline{\Delta}(s) = \prod_{i=1}^{n} (s - \lambda_i) = s^n + \alpha_1' s^{n-1} + \cdots + \alpha_{n-1}' s + \alpha_n'$.

(2) Compute $\Delta(s) = s^n + \alpha_1 s^{n-1} + \cdots + \alpha_{n-1} s + \alpha_n$, which is the characteristic polynomial of $\mathcal{A}$.

(3) Compute $\overline{\mathcal{A}} = \begin{bmatrix} -\alpha_1' & \cdots & -\alpha_{n-1}' & -\alpha_n' \\ 1 & & 0 & 0 \\ 0 & \ddots & 0 & 0 \\ 0 & 0 & 1 & 0 \end{bmatrix}$, $\overline{\mathcal{B}} = \begin{bmatrix} 1 \\ 0 \\ 0 \\ 0 \end{bmatrix}$.

(4) Compute $\overline{\mathcal{C}} = \begin{bmatrix} \mathcal{B} \ \mathcal{AB} \cdots \mathcal{A}^{n-1}\mathcal{B} \end{bmatrix}$ and $\overline{\mathcal{C}}' = \begin{bmatrix} \overline{\mathcal{B}} \ \overline{\mathcal{AB}} \cdots \overline{\mathcal{A}}^{n-1}\overline{\mathcal{B}} \end{bmatrix}$.

(5) $\mathcal{K} = \begin{bmatrix} \alpha_1' - \alpha_1 & \cdots & \alpha_n' - \alpha_n \end{bmatrix} \overline{\mathcal{C}}' \overline{\mathcal{C}}^{-1}$.

Similarly, Ackermann's formula, uses the following steps to place the poles:

(1) Compute $\overline{\Delta}(s) = \prod_{i=1}^{n} (s - \lambda_i) = s^n + \alpha_1' s^{n-1} + \cdots + \alpha_{n-1}' s + \alpha_n'$.

(2) Compute $\Delta(s) = s^n + \alpha_1 s^{n-1} + \cdots + \alpha_{n-1} s + \alpha_n$, which is the characteristic polynomial of $\mathcal{A}$.

(3) Compute $\overline{\mathcal{A}} = \begin{bmatrix} -\alpha_1' & \cdots & -\alpha_{n-1}' & -\alpha_n' \\ 1 & & 0 & 0 \\ 0 & \ddots & 0 & 0 \\ 0 & 0 & 1 & 0 \end{bmatrix}$, $\overline{\mathcal{B}} = \begin{bmatrix} 1 \\ 0 \\ 0 \\ 0 \end{bmatrix}$.

(4) Compute $\overline{\mathcal{C}} = \begin{bmatrix} \mathcal{B} \ \mathcal{AB} \cdots \mathcal{A}^{n-1}\mathcal{B} \end{bmatrix}$ and $\overline{\mathcal{C}}' = \begin{bmatrix} \overline{\mathcal{B}} \ \overline{\mathcal{AB}} \cdots \overline{\mathcal{A}}^{n-1}\overline{\mathcal{B}} \end{bmatrix}$.

(5) $\mathcal{K} = \begin{bmatrix} 0 \cdots 01 \end{bmatrix}_{1\times n} \overline{\mathcal{C}}^{-1} \overline{\Delta}(\mathcal{A})$.

It is also possible to design the observer using Ackermann's formula which gives

$$\mathcal{L} = \overline{\Delta}(\mathcal{A}) \mathcal{O}^{-1} \begin{bmatrix} 0 \cdots 01 \end{bmatrix}_{1\times n}^{T}. \tag{5.27}$$

When using pole placement, the following points should be borne in mind:

(1) The magnitude of $\mathcal{K}$ gives the amount of effort required to control the process. The further the desired poles are from the actual poles of the system, the larger the controller gain $\mathcal{K}$.
(2) For multi-input systems, $\mathcal{K}$ is not unique.
(3) $(\mathcal{A}, \mathcal{B})$ and $(\mathcal{A} - \mathcal{B}\mathcal{K}, \mathcal{B})$ are controllable. However, due to pole-zero cancellations, the resulting system may not be observable.
(4) Discrete systems can be controlled in the same manner, *mutatis mutandi.*

Tuning a PID Controller

When designing a PID controller, there exist different approaches that can be taken to specifying the initial controller parameters. In practice, once these initial parameters have been obtained, they will be fine-tuned to obtain the desired performance. There are two main approaches to controller tuning: **model-based** and **structure-based**.

In model-based controller tuning, a model of the process is required and the closed-loop behaviour of the resulting system is specified. Using this information, it is possible to obtain the resulting form of the controller transfer function. Comparing the resulting controller transfer function with the standard PID controller allows equations for the constants to be determined. The most common model-based approach is the internal model control (IMC) framework. The main advantage of this approach is that it allows the engineer to specify the desired closed-loop behaviour of the system, while the main disadvantage of this approach is that a relatively accurate model of the system is required. Note that the specification of the closed-loop transfer function can be used to add robustness to the system by changing the speed (time constant) of the response. The faster the system responds to changes the less robust is the resulting closed-loop system. In many cases, the simplified IMC (SIMC) rules provide better control performance for industrial systems. The SIMC rules take into consideration additional factors, such as disturbances in the input and faster response for systems with large time constants. SIMC rules recommend PI controllers for first-order systems and PID controllers for second-order systems.

In structure-based controller tuning, the structure (or type) of a controller, the objective (disturbance rejection or setpoint tracking), and metric for measuring good control are specified. Minimising this metric allows the values of the parameters to be determined. The most common structure-based approach is using the integrated time-averaged error (ITAE) as the metric. Using the ITAE metric penalises persistent errors, which can be small errors that last for a long period of time, and provides a conservative approach to control. The main advantage of this method is that a model of the system need not be provided, while the main disadvantages are that the engineer has no control over the closed-loop response of the system and that the resulting system is not very robust to changes in the process.

Table 5.1 PI controller constants for first-order-plus-deadtime models

Controller method		K_c	τ_I
Structure-based	ITAE (setpoint tracking)	$\frac{0.586}{K}\left(\frac{\theta}{\tau}\right)^{-0.916}$	$\frac{\tau}{1.03-0.165\left(\frac{\theta}{\tau}\right)}$
	ITAE (disturbance rejection)	$\frac{0.859}{K}\left(\frac{\theta}{\tau}\right)^{-0.977}$	$\frac{\tau}{0.674}\left(\frac{\theta}{\tau}\right)^{0.680}$
Model-based	SIMC (τ_c chosen so that τ_c>0.8θ and τ_c>0.1τ)	$\frac{1}{K}\left(\frac{\tau}{\theta+\tau_c}\right)$	$\min\left(\tau, 4(\tau_c+\theta)\right)$

Table 5.2 PID controller constants for first-order-plus-deadtime models

Controller method		K_c	τ_I	τ_D
Structure-based	ITAE (setpoint tracking)	$\frac{0.965}{K}\left(\frac{\theta}{\tau}\right)^{-0.85}$	$\frac{\tau}{0.796-0.147\left(\frac{\theta}{\tau}\right)}$	$0.308\tau\left(\frac{\theta}{\tau}\right)^{0.929}$
	ITAE (disturbance rejection)	$\frac{1.357}{K}\left(\frac{\theta}{\tau}\right)^{-0.947}$	$\frac{\tau}{0.842}\left(\frac{\theta}{\tau}\right)^{0.738}$	$0.381\tau\left(\frac{\theta}{\tau}\right)^{0.995}$
Model-based	IMC (τ_c chosen so that τ_c>0.8θ and τ_c>0.1τ)	$\frac{1}{K}\left(\frac{2\frac{\tau}{\theta}+1}{2\frac{\tau_c}{\theta}+1}\right)$	$\frac{\theta}{2}+\tau$	$\frac{\tau}{2\left(\frac{\tau}{\theta}\right)+1}$

Table 5.3 PID controller constants for second-order-plus-deadtime models

Controller method		$\tilde{K}_c$	$\tilde{\tau}_I$	$\tilde{\tau}_D$
Model-based	SIMC $G_p = \dfrac{K}{(\tau_1 s+1)(\tau_2 s+1)}e^{-\theta s}$ with $\tau_1 > \tau_2$	$\frac{1}{K}\left(\frac{\tau_1}{\theta+\tau_c}\right)$	$\min\left(\tau_1, 4(\theta+\tau_c)\right)$	τ_2

Tables 5.1 and 5.2 present some common tuning methods for first-order transfer functions. Table 5.3 presents the SIMC tuning method for a second-order transfer function. It should be noted that the PID controller has the following series form:

$$G_c = \tilde{K}_c\left(1+\frac{1}{\tilde{\tau}_I s}\right)(1+\tilde{\tau}_D s) \tag{5.28}$$

where the controller parameters with the tilde $\tilde{\circ}$ have the same meaning as the corresponding controller parameters without a tilde.

Example 5.6: Designing a PI Controller

Consider the following first-order system

$$G_p = \frac{1.5}{20s+1}e^{-10s} \tag{5.29}$$

Determine the controller parameters for a PI controller using the SIMC method.

Solution

Before solving the problem, it is helpful to determine the values of the various constants:

Gain: $K=1.5$

Time Constant: $\tau=20$

Time Delay: $\theta=10$

For the SIMC method, there is a need to also specify the value of τ_c, which is essentially the closed-loop time constant. The smaller the value, the faster the response is, but less robust is the overall system. The constraints given with this method imply that $\tau_c>0.8\theta=8$ and $\tau_c>0.1\tau=2$. For the purposes of this example, set the value of τ_c to be 10, which satisfies both constraints.

Therefore, the controller parameters can be computed from Table 5.1 as follows:

$$K_c = \frac{1}{K}\frac{\tau}{\theta+\tau_c} = \frac{1}{1.5}\left(\frac{20}{10+10}\right) = \frac{2}{3} \tag{5.30}$$

$$\tau_I = \min\left(\tau, 4(\tau_c+\theta)\right) = \min\left(20, 4(10+10)\right) = 20 \tag{5.31}$$

The resulting behaviour of the closed-loop system is shown in Fig. 5.13. It can be seen that the process responds almost exactly as expected.

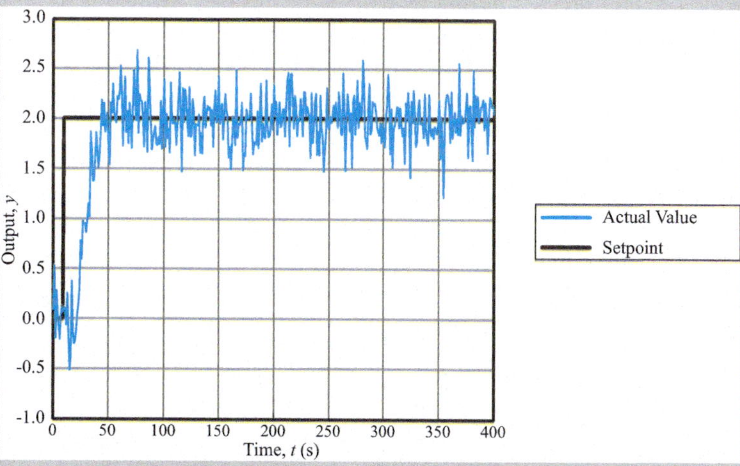

Fig. 5.13 Closed-loop performance of the PI controller

Control Performance

Once a controller has been designed, it is often necessary to measure its performance and determine if the controller is attaining its objectives. As with any such measurements, performance is measured against some (engineering-related) specification. Unlike most other equipment, the specification for a controller involves time, either explicitly when speed of response is a criterion, or implicitly, for statistics such as standard deviation. The performance of a controller can be measured based on two different sets of criteria: how the controller responds to a change in setpoint, that is, setpoint tracking or the servoresponse, and how the controller responds to disturbances, that is, the regulatory response.

Since the objective in setpoint tracking is how well the controller responds to a setpoint change, it is easy to determine an appropriate measure for performance assessment. For this reason, setpoint tracking is often used to design a controller. As a first approximation, a closed-loop system can be considered to be a second-order system with time delay. Therefore, the response of the closed-loop system to a step change in the setpoint would be the step response of a second-order system. The most common performance measures for setpoint tracking are just characteristics of a second-order system responding to a step input, that is, rise time, overshoot, and settling time of the controlled variable, as shown in Fig. 5.14. The **rise time**, τ_r, is defined as the first time the process reaches the setpoint value. The **overshoot** is defined as the ratio of how much the process goes over (or under) the setpoint divided by the magnitude of the step change. Note that if the step change is negative, then the overshoot will also be negative (that is, it is really an undershoot) so that the ratio remains positive. The **settling time**, τ_s, is defined as the last time for which the process lies outside the 5% bounds. The 5% bounds are defined as a boundary on either side of the new setpoint that is defined as 2.5% of the difference $(y_{ss,2} - y_{ss,1})$, where y_{ss} is the steady-state value and the numeric subscripts represent the initial and final values. The settling time is roughly three times the closed-loop time constant.

On the other hand, for performance assessment of the regulatory response, the objective is less certain. Instead, various quantitative measures are used to quantify performance assessment, including:

(1) **Standard deviation of the manipulated variable**: This is the simplest and most important measure of control performance. However, it is difficult to specify the desired value *a priori* since it depends on the nature and magnitude of the disturbances affecting the process. What can be specified, or at least described qualitatively, is the trade-off between control error and control action. One way of thinking of a controller is that it moves variation from one variable (usually the controlled variable) to another (usually the manipulated variable). As disturbances affect the process, the controller will respond by changing the manipulated variable. The response will then mitigate the effect of the disturbances. Larger disturbances will require greater control action than small disturbances.

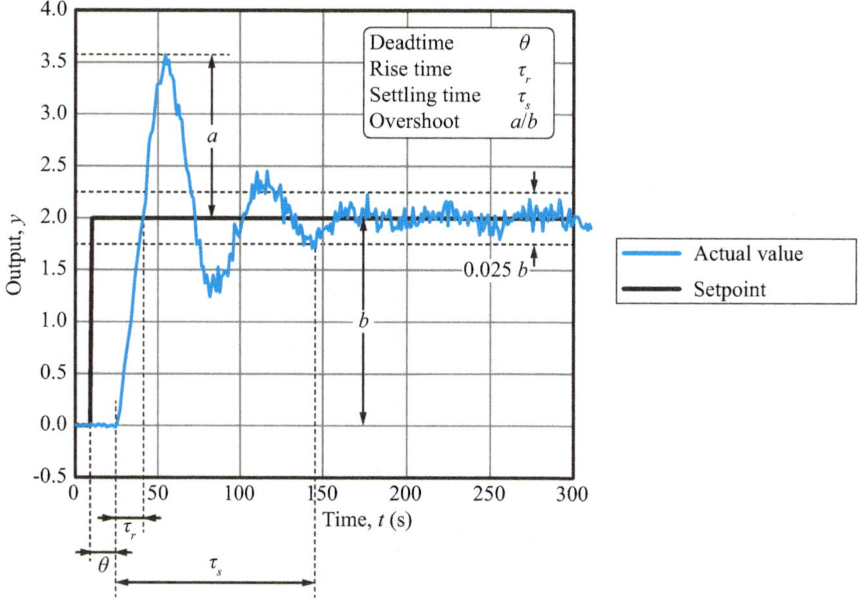

Fig. 5.14 Measures for assessing the performance of a controller for setpoint tracking

(2) **Mean of the output variable**: This can be used to assess how well the controller maintains the given setpoint in the presence of disturbances.

(3) **Advanced performance assessment**: More advanced performance assessment methods are available, such as the minimum variance controller and the Harris Index (Shardt, et al., 2012), which can provide appropriate benchmarks for determining good performance.

5.2 Feedforward Control

In feedforward control, a measurable disturbance is used as the input to a controller, so as to take corrective action before the disturbance affects the process. The goal when using feedforward control is to design a controller that can correct the disturbance at the same time as the disturbance affects the process. In practice, such a controller can rarely be designed, since there will be time delay in the system, that is, there is some time period between when the variable is measured and when it affects the system. If the time delay between the disturbance and the process is smaller than the time delay between the input and the process, then it is not possible to compensate the disturbance before it has affected the system. Figure 5.15 shows a typical feedforward control loop.

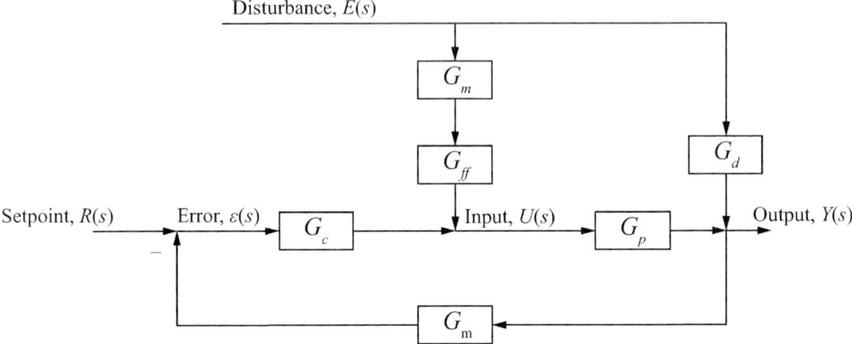

Fig. 5.15 Block diagram for feedforward control

A special type of feedforward controller is a **decoupler**, which seeks to separate two or more interacting processes so that they can be treated as independent systems controlled by their own SISO control loops. A decoupler basically takes the controller output from the other controllers and treats it as a measurable disturbance that needs to be counteracted. Decouples can be very effective in controlling a multivariate system when the interactions are relatively straight forward and not too nonlinear.

In feedforward control design, the objective to design a controller that can compensate for a measured disturbance before it can affect the process. Assume that the process model is G_p and the disturbance model is G_d, then for the feedforward controller shown in Fig. 5.15, the basic form of the feedforward controller can be written as

$$G_{ff} = -\frac{G_d}{G_p}. \tag{5.32}$$

In the ideal situation, this will exactly cancel out the effect of the measured disturbance. However, in practice, such a controller cannot be implemented for the following two reasons:

(1) **Time Delay**: If the time delay in the process response is larger than the time delay in the disturbance response, then the overall time delay for the process will be negative requiring knowledge of future information about the process. Since such a situation is impossible, the general solution is to drop the unrealisable time delay from the final controller.

(2) **Unstable Zeroes in G_p**: If the process contains unstable zeroes, then the resulting transfer function will also be unstable. In such a case, it will not be possible to realise the controller. A common solution in this case is to drop the offending zeroes from the final representation.

There are two common feedforward controllers that can be designed:

(1) **Static controllers** where only the gains of the process are considered, that is,

$$G_{ff} = -\frac{K_d}{K_p} \tag{5.33}$$

where K_d is the disturbance gain and K_p is the process gain.

(2) **Dynamic controllers** where the process and disturbance dynamics are included in the final control law. Very often, the controller takes the form of a lead-lag controller, that is,

$$G_{ff} = -\frac{K_d\left(\tau_p s + 1\right)}{K_p(\tau_d s + 1)} e^{-\theta_{ff} s} \tag{5.34}$$

where τ_p is the process time constant, τ_d is the disturbance time constant, and $\theta_{ff} = \theta_d - \theta_p$. If θ_{ff} is negative, then the term is often ignored. Ideally, $\left|\tau_p/\tau_d\right| < 1$, which avoids large spikes when this controller is used.

Example 5.7: Designing a Feedforward Controller
Given the following system, design both a dynamic and a static feedforward controller. Compare the performance of both. The system parameters are:

$$G_p = \frac{-1.5(s-1)}{(s+1)(10s+1)} e^{-10s} \tag{5.35}$$

$$G_d = \frac{-0.5}{(5s+1)(7s+1)} e^{-5s} \tag{5.36}$$

$$G_c = \frac{2}{3}\left(1 + \frac{1}{20s}\right). \tag{5.37}$$

For the simulation, assume that the measured disturbance is driven by Gaussian white noise with a magnitude of 0.5 and there is a step change in the setpoint of 2 units after 100 s.

Solution

Using Eq. (5.33), the static feedforward controller can be computed as

$$G_{ff,s} = -\frac{K_d}{K_p} = -\frac{-0.5}{1.5} = \frac{1}{3}. \tag{5.38}$$

Note that the gain can be obtained by setting $s=0$ in the given transfer functions and evaluating the resulting value.

When creating the dynamic feedforward controller, it is important that the transfer functions be written in their standard form, that is, all roots are of the form $\tau s+1$. In this particular example, the zero of the process transfer function needs to be written into this form to give

$$G_p = \frac{1.5(-s+1)}{(s+1)(10s+1)}e^{-10s}. \tag{5.39}$$

The dynamic feedforward controller can be designed using Eq. (5.32). This gives

$$G_{ff,d} = -\frac{G_d}{G_p} = -\frac{\frac{-0.5}{(5s+1)(7s+1)}e^{-5s}}{\frac{1.5(-s+1)}{(s+1)(10s+1)}e^{-10s}} = \frac{1}{3}\frac{(s+1)(10s+1)}{(5s+1)(7s+1)(-s+1)}e^{5s}. \tag{5.40}$$

However, this controller as written cannot be realised since there is an unstable pole and the time delay requires future values. In order to obtain a realisable dynamic feedforward controller, both the unstable pole and time delay terms will be dropped. This gives the following form for the controller

$$G_{ff,d} = \frac{1}{3}\frac{(s+1)(10s+1)}{(5s+1)(7s+1)}. \tag{5.41}$$

Figure 5.16 shows the effects of both feedforward controllers on the process. It can be seen that without the feedforward controller the process has greater oscillations. Adding the static feedforward controller tends to decrease the effect of the oscillations. However, the behaviour is still rather jittery. The dynamic feedforward controller has smoothed out the jitter and the behaviour is cleaner.

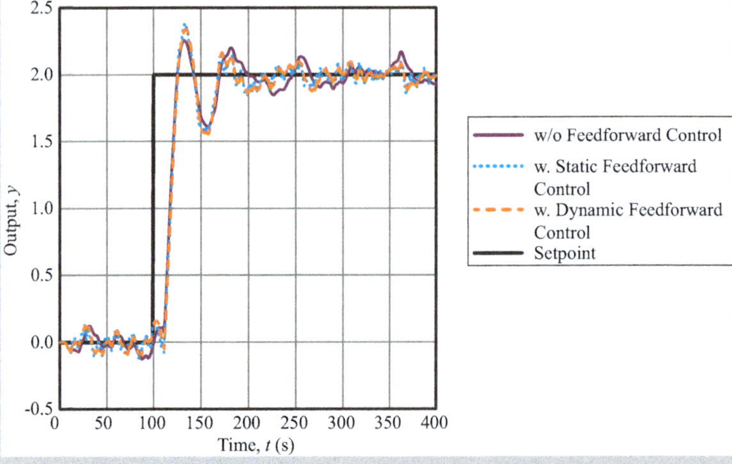

Fig. 5.16 Effect of feedforward control on a process

5.3 Discrete-Event Control

In discrete-event control, a specific event or series of events triggers a given control sequence. This can often be seen when dealing with safety-oriented control situations, where, for example, a pressure relief valve will only open if the pressure is above a certain value and will close once the pressure falls below the threshold.

Another type of discrete-event control is **interlocking**, where a specified set of circumstances must hold before a given action can occur, for example, a microwave cannot be started until the microwave door has been closed, which minimises the effect of the radiation on the user. Interlocking is an important component of safety considerations in order to prevent operators from taking inappropriate action. Designing the correct interlocks can be a challenging proposition.

A special type of discrete-event control occurs when binary signals are used, for example, a sensor will report when the elevator has reached the correct floor, and then the motor of the drive will be switched off. The exact height of the elevator above the ground is not continuously measured and monitored. When dealing with binary signals, there are two possible types of controllers: **logic control** and **sequential control**.

In logic controllers, also called combinatorial controllers, one obtains the control variable as a combination, that is, a logical operation, of signals, for example, system output signals (sensor values such as door open or closed) or user inputs (*e.g.* emergency button pressed or not pressed). Consider a simple example of a lathe. A lathe is turned on when the user presses the power button *and* the chuck is closed. The lathe is switched off when an end position is reached *or* the emergency stop button is pressed.

In sequential control, individual control operations are performed in certain steps. The advancement to the next step occurs either after a certain time has passed (time-dependent sequential control), for example, a simple traffic light, or in the presence of a specific event (process-dependent sequential control), for example, traffic lights that change their signal when a pedestrian presses the button. In contrast to the logic controllers, it is therefore not possible to clearly determine the manipulated variable even if all signal values are known, since one also has to know at which point in the process, one is currently located.

5.4 Supervisory Control

Supervisory control focuses on designing controllers that can deal with complex, interacting systems incorporating physical, process, and economic constraints, as well as dealing with changing economic circumstances. Supervisory control systems will often set the setpoint for **subsidiary** (or **slave**) controllers. They can be used to control large systems containing many different variables and even locations. A supervisory control system can range from a simple nested control-loop system to complex model predictive control systems. There are two primary supervisory control strategies: **cascade control** and **model predictive control (MPC)**, which is often called advanced process control (APC) in the petrochemical industry.

5.4.1 Cascade Control

In cascade control, at least two control loops are nested within each other. A typical cascade-control strategy is shown in Fig. 5.17, where it is desired to ensure that the flow rate controlled by the valve is as accurate as possible, so that the process experiences the tightest control possible. The **primary** or **master** control loop is the outer control loop, which sets the setpoint for the **secondary** or **slave** control loop. The general procedure for tuning a cascade-control loop is to first tune the innermost loop and then work up towards the uppermost loop, treating all tuned controllers as part of the process. The closed-loop time constant of the inner loop should be smaller (faster) than the closed-loop time constant of the outer loop. A common rule of thumb is that the ratio of closed-loop time constant of the primary loop and that of the secondary loop ($\tau_{c,p}/\tau_{c,s}$) should be between 4 and 10. Too small a ratio implies that the secondary loop will not have reached steady state and there will be interactions between the two loops leading to a loss of overall performance. Too large a ratio implies that the advantages of cascade control will be lost as it will slow down the overall response of the system. However, it will make it more robust to changes in the underlying process models.

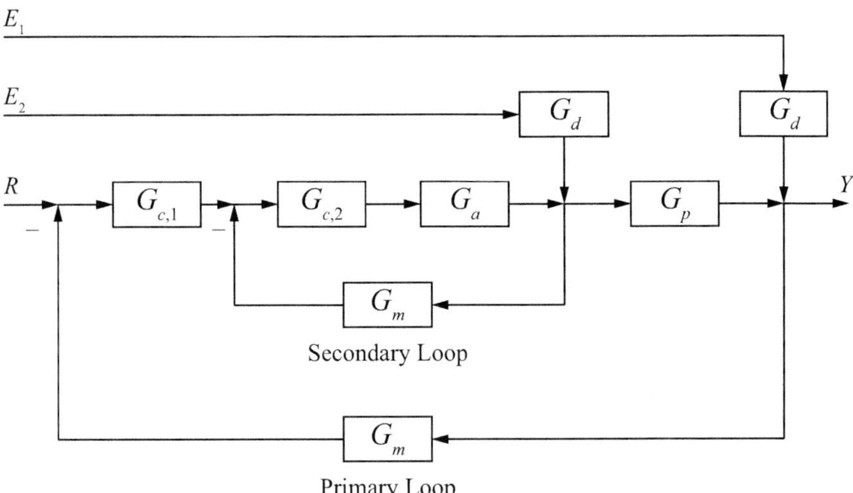

Fig. 5.17 Block diagram for cascade control

5.4.2 Model Predictive Control

Model predictive control is an advanced control strategy that can take into consideration not only the deviations from the setpoint, but also various economic and physical constraints, for example, it can explicitly handle the situation where the water in the tank should not go above a given value. In order to accurately handle these constraints, this approach requires a good model and proper engineering design. As shown in Fig. 5.18, model predictive control works by optimising the control actions over a period of time, often called the **control horizon**, using the predicted values from the model. The **prediction horizon** is the period over which the model is used to predict the behaviour of the system. After the prediction horizon has been reached, it is assumed that the process is at steady state and will not change its values. Any constraints are then implemented as necessary on the predicted results. The controller then implements the next control action and repeats the optimisation procedure again. This ensures that the effect of disturbances, plant-model mismatch, and other imperfections in the system do not have too large an effect on the overall performance.

Model predictive control is effective in handling complex process with multiple inputs and outputs that interact strongly. Although the original version was designed for linear systems, variants exist for nonlinear systems.

One of the key disadvantages of model predictive control is that it requires a good model that must be updated as necessary. A poor model can cause the control system to degrade and forgo the benefits of the approach.

There are many different implementations of model predictive control. The most popular implementation in the industry is the dynamic-matrix controller (DMC) approach, which is presented here. The objective function for model predictive control is

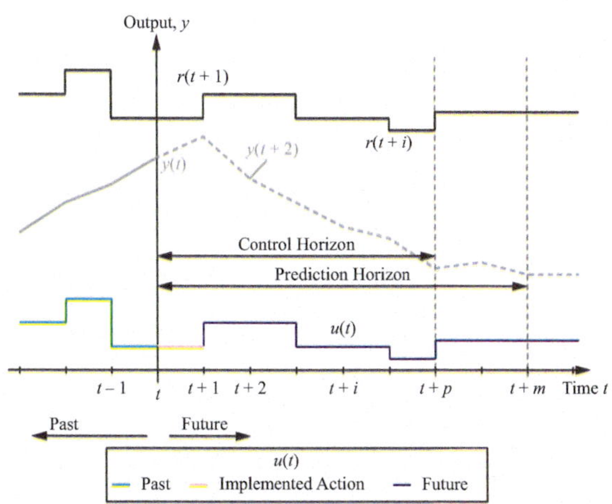

Fig. 5.18 Model predictive control

$$\min_{\Delta\vec{u}} \left[\left(\vec{r} - \hat{\vec{y}}\right)^T \mathcal{Q} \left(\vec{r} - \hat{\vec{y}}\right) + (\Delta\vec{u})^T \mathcal{R} (\Delta\vec{u}) \right]$$
$$\text{subject to} \tag{5.42}$$
$$\hat{\vec{y}} = \vec{y}^* + \mathcal{A}\Delta\vec{u}$$

where $\vec{r}$ is the reference vector, $\hat{\vec{y}}$ the vector of the predicted process values with control, $\mathcal{Q}$ the process scaling matrix, $\Delta\vec{u}$ the vector of the changes in the controller action, $\mathcal{R}$ the input scaling matrix, $\vec{y}^*$ the vector of the process values without control, and $\mathcal{A}$ the dynamic matrix. Furthermore, let m be the control horizon, p the prediction horizon, d the process time delay, and n the settling time. It should be noted that $1 \leq m \leq p\text{–}d$ and $p > d$.

The solution requires the step response of the process, that is,

$$y_t = \sum_{i=1}^{\infty} a_i z^{-i} \Delta u_t \tag{5.43}$$

where a_i is the step-response coefficient. The step-response coefficient can be calculated either by polynomial division of the transfer function model or using the coefficients of the impulse response. The coefficients a_i of the step response have the following relationship to the coefficients of the impulse response h_j

$$a_i = \sum_{j=1}^{i} h_j. \tag{5.44}$$

Furthermore, the settling time is defined as the time at which the first coefficient a_n is within a range of between 0.975 and 1.025 times the value of a_∞, where a_∞ is the steady-state value. All step-response coefficients after the settling time n can be assumed to be equal to a_{n+1}. The $p \times m$ dynamic $\mathcal{A}$ matrix can written as

$$\mathcal{A} = \begin{bmatrix} a_1 & 0 & \cdots & 0 \\ a_2 & a_1 & \ddots & \vdots \\ \vdots & \vdots & \ddots & 0 \\ \vdots & \vdots & & a_1 \\ a_p & a_{p-1} & \cdots & a_{p-m+1} \end{bmatrix}. \tag{5.45}$$

For a univariate system with only one input variable and one output variable, the solution for Eq. (5.42) is

$$\Delta\vec{u} = \vec{K}_C \vec{e} = \left(\mathcal{A}^T \mathcal{Q} \mathcal{A} + \mathcal{R}\right)^{-1} \mathcal{A}^T \mathcal{Q}^T \left(\vec{r} - \vec{y}^*\right) \tag{5.46}$$

where

$$y^*_{t+l} = y_t - \sum_{i=1}^{n} (a_i - a_{l+i})\Delta u_{t-i}. \tag{5.47}$$

Once the solution has been found, then the first control action Δu_1 will be implemented by the controller. At the next sampling time, the above optimisation procedure is repeated, a new optimal value is calculated, and the first control action is implemented. This allows the system to take unexpected process changes into consideration.

Example 5.8: Design of a Model Predictive Controller
Design a model predictive controller for the following SISO system:

$$G_p = \frac{2z^{-1}}{1 - 0.75z^{-1}}. \tag{5.48}$$

Set up the required matrices and perform the first step of the iteration. For this, assume that $m = p = 3$. Let $\mathcal{Q} = \mathcal{I}_p$ and $\mathcal{R} = \mathcal{J}_3$, where $\mathcal{J}_n$ is the $n \times n$ identity matrix. Furthermore, a step change occurs at $t = 0$, with the process having previously been in steady state.

Solution

To determine the model predictive controller, we must first determine the step-response model. This can be determined using long division, which gives the impulse-response coefficients from which the step-response coefficients can be easily calculated. Thus,

$$\begin{array}{r l}
 & 2z^{-1} + 1.5z^{-2} + 1.125z^{-3} \\
1 - 0.75z^{-1} & \overline{)2z^{-1}} \\
 & \underline{-2z^{-1} + 1.5z^{-2}} \\
 & 1.5z^{-2} \\
 & \underline{-1.5z^{-2} + 1.125z^{-3}} \\
 & 1.125z^{-3}\ldots
\end{array} \tag{5.49}$$

It can be seen that $h_i = 2(0.75)^{i-1}$ for $i \geq 1$. In general, for a transfer function of the form

$$G_p = \frac{\beta z^{-d}}{1 - \alpha z^{-1}} \tag{5.50}$$

the coefficients of the impulse response can be given as

$$h_i = \begin{cases} \beta\alpha^{i-d} & i \geq d \\ 0 & i < d \end{cases}. \tag{5.51}$$

The required step-response coefficients can be found using Eq. (5.44)

$$a_i = \sum_{j=0}^{i} h_j = \sum_{j=1}^{i} 2\left(0.75^{j-1}\right) \tag{5.52}$$

Equation (5.52) gives a geometric series that can be written as

$$a_i = \beta\left(\frac{1-\alpha^{i-d+1}}{1-\alpha}\right) = 2\frac{\left(1-0.75^{i}\right)}{1-0.75}. \tag{5.53}$$

It follows that

$$\begin{aligned} a_1 &= 2\frac{(1-0.75)}{1-0.75} = 2 \\ a_2 &= 2\frac{\left(1-0.75^{2}\right)}{1-0.75} = 3.5 \\ a_3 &= 2\frac{\left(1-0.75^{3}\right)}{1-0.75} = 4.625. \end{aligned} \tag{5.54}$$

Since $d = 1$ for this example, the step-response coefficients with a smaller d have a value of zero and can thus be ignored in the summation.

In the next step, the settling time is calculated. For a converging geometric series (implies that the process of interest is stable), the value to which the series converges is given by

$$K_p = a_\infty = \frac{\beta}{1-\alpha} = \frac{2}{1-0.75} = 8. \tag{5.55}$$

This gives the settling time for the first value for which the process lies in the interval 0.975×8 and 1.025×8. Since the process under consideration has no oscillations, only the lower limit is of interest to us. Given that the settling time is equal to the value of i in Eq. (5.53), we get

$$\beta\left(\frac{1-\alpha^{n-d+1}}{1-\alpha}\right) = \frac{0.975\beta}{1-\alpha}, \tag{5.56}$$

which can be solved for n to give

$$n = \frac{\ln 0.025}{\ln \alpha} + d - 1. \tag{5.57}$$

In our case, we obtain

$$n = \frac{\ln 0.025}{\ln 0.75} + 1 - 1 = 12.8 = 13. \tag{5.58}$$

Note that n is always an integer, so we need to round the resulting value up to the nearest integer value.

The 3×3 dynamic matrix $\mathcal{A}$ is then

$$\mathcal{A}=\begin{bmatrix} a_1 & 0 & \cdots & 0 \\ a_2 & a_1 & \ddots & \vdots \\ \vdots & \vdots & \ddots & 0 \\ \vdots & \vdots & & a_1 \\ a_p & a_{p-1} & \cdots & a_{p-m+1} \end{bmatrix}=\begin{bmatrix} 2 & 0 & 0 \\ 3.5 & 2 & 0 \\ 4.625 & 3.5 & 2 \end{bmatrix}. \tag{5.59}$$

This gives

$$\begin{aligned}\mathcal{A}^T\mathcal{Q}\mathcal{A}+\mathcal{R}&=\begin{bmatrix} 2 & 0 & 0 \\ 3.5 & 2 & 0 \\ 4.625 & 3.5 & 2 \end{bmatrix}^T\begin{bmatrix} 1 & 0 & 0 \\ 0 & 1 & 0 \\ 0 & 0 & 1 \end{bmatrix}\begin{bmatrix} 2 & 0 & 0 \\ 3.5 & 2 & 0 \\ 4.625 & 3.5 & 2 \end{bmatrix}+\begin{bmatrix} 1 & 0 & 0 \\ 0 & 1 & 0 \\ 0 & 0 & 1 \end{bmatrix}\\ &=\begin{bmatrix} 38.640\,625 & 23.1875 & 9.25 \\ 23.1875 & 17.25 & 7 \\ 9.25 & 7 & 5 \end{bmatrix}.\end{aligned} \tag{5.60}$$

The required inverse is then

$$\begin{aligned}\left(\mathcal{A}^T\mathcal{Q}\mathcal{A}+\mathcal{R}\right)^{-1}&=\begin{bmatrix} 38.640\,625 & 23.1875 & 9.25 \\ 23.1875 & 17.25 & 7 \\ 9.25 & 7 & 5 \end{bmatrix}^{-1}\\ &=\begin{bmatrix} 0.134046 & -0.184\,200 & 0.009\,896 \\ -0.184200 & 0.387\,349 & -0.201518 \\ 0.009\,896 & -0.201\,518 & 0.463\,818 \end{bmatrix}.\end{aligned} \tag{5.61}$$

For the controller gain, we obtain the result

$$\begin{aligned}\vec{K}_C&=\left(\mathcal{A}^T\mathcal{Q}\mathcal{A}+\mathcal{R}\right)^{-1}\mathcal{A}^T\mathcal{Q}^T\\ &=\begin{bmatrix} 0.134046 & -0.184\,200 & 0.009\,896 \\ -0.184200 & 0.387\,349 & -0.201518 \\ 0.009\,896 & -0.201\,518 & 0.463\,818 \end{bmatrix}\begin{bmatrix} 2 & 0 & 0 \\ 3.5 & 2 & 0 \\ 4.625 & 3.5 & 2 \end{bmatrix}^T\begin{bmatrix} 1 & 0 & 0 \\ 0 & 1 & 0 \\ 0 & 0 & 1 \end{bmatrix}\\ &=\begin{bmatrix} 0.268\,091 & 0.100\,759 & -0.004\,948 \\ -0.368\,400 & 0.129\,997 & 0.100\,759 \\ 0.019\,792 & -0.368\,400 & 0.268\,091 \end{bmatrix}.\end{aligned} \tag{5.62}$$

The reference signal for the next three sample periods is

$$\vec{r} = \begin{bmatrix} 1 \\ 1 \\ 1 \end{bmatrix}. \tag{5.63}$$

The predicted uncontrolled position will be equal to the steady-state value, *i.e.* zero, since no control has been made. This implies that

$$\vec{y}^* = \begin{bmatrix} 0 \\ 0 \\ 0 \end{bmatrix}. \tag{5.64}$$

The controller action is then

$$\begin{aligned} \Delta\vec{u} &= \vec{K}_C\vec{e} = \vec{K}_C\left(\vec{r} - \vec{y}^*\right) \\ &= \begin{bmatrix} 0.268\,091 & 0.100\,759 & -0.004\,948 \\ -0.368\,400 & 0.129\,997 & 0.00\,759 \\ 0.019\,792 & -0.368\,400 & 0.268\,091 \end{bmatrix} \begin{bmatrix} 1-0 \\ 1-0 \\ 1-0 \end{bmatrix} \\ &= \begin{bmatrix} 0.363\,902 \\ -0.137\,644 \\ -0.080\,517 \end{bmatrix}. \end{aligned} \tag{5.65}$$

Only the first controller action $\Delta u_1 = 0.363\,902$ will be implemented. At the next time point, Eq. (5.65) will be recomputed using the new values for r and y^*.

For a multivariate system, the vectors and matrices become *supervectors* and *supermatrices*, meaning that a vector consists of many vectors, for example

$$\Delta\vec{u} = \begin{bmatrix} \Delta\vec{u}_1 \\ \vdots \\ \Delta\vec{u}_h \end{bmatrix} \tag{5.66}$$

where $\Delta\vec{u}_i$ is the input vector for the ith input. Let us consider a MIMO system with s outputs and h inputs. This gives a dynamic matrix $\mathcal{A}$ with the following form

$$\mathcal{A} = \begin{bmatrix} \mathcal{A}_{11} & \cdots & \mathcal{A}_{1h} \\ \vdots & & \vdots \\ \mathcal{A}_{s1} & \cdots & \mathcal{A}_{sh} \end{bmatrix} \tag{5.67}$$

where $\mathcal{A}_{ij}$ is the dynamic matrix between the jth input and ith output. From Eq. (5.46), it follows that

$$y^*_{i,t+l} = y_{i,t} - \sum_{j=1}^{h} \sum_{k=1}^{n} \left(a_{ij,k} - a_{ij,l+k}\right) \Delta u_{j,t-k} \tag{5.68}$$

and

$$\Delta u_j = \sum_{i=1}^{ps} k_{ji}\left(r_i - y_i^*\right), \quad j = km + 1,\ k = 0, 1, ..., h - 1 \tag{5.69}$$

where $a_{ij,\ k}$ is the kth coefficient of the step response for the process between the jth input and ith output and $\Delta u_{j,\ t-k}$ is the change in the control action for the jth input at time point $t-k$.

5.5 Advanced Control Strategies

In addition to the control strategies mentioned so far, there exist various useful strategies that can be combined with the aforementioned approach to obtain a better result. Such methods include the **Smith predictor**, **deadbanding**, **squared control**, **ratio control**, **input-position control**, and **characterisation of nonlinearities**.

5.5.1 *Smith Predictor*

The Smith predictor is an approach that seeks to minimise the effect of time delay on the control strategy. It is useful when there is a large time delay in the system that needs to be dealt with. The control strategy for this approach is shown in Fig. 5.19.

5.5.2 *Deadbanding and Gain Scheduling*

Deadbanding refers to the idea that when the control error is within some band of the reference point, then there is no need to perform additional control. Deadbanding means that as long as the value is within the band then no control

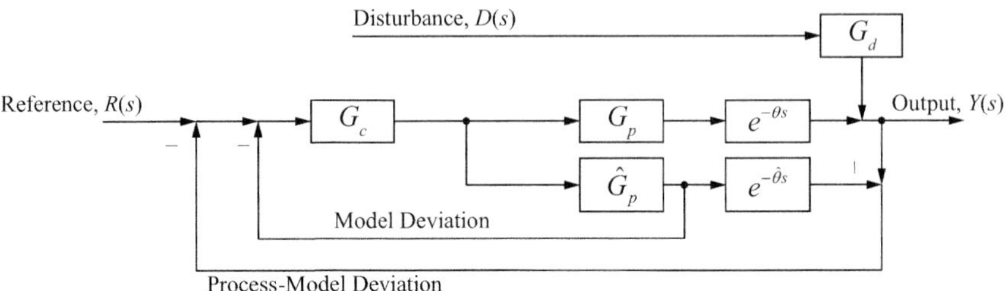

Fig. 5.19 Block diagram for control using a Smith predictor

action will be made. In turn, once the error strays outside the deadband, then control will be performed. Mathematically, this can be written as

$$u_t = \begin{cases} 0 & |\varepsilon_t| < K_{db} \\ G_c & \text{otherwise} \end{cases} \tag{5.70}$$

where K_{db} is the deadband constant. The deadband can be specified as either some fixed value of the setpoint (for example, within 5°C of the setpoint) or as a percentage (within 5% of the setpoint). Deadbanding is useful when tight control is not desired, that is, small deviations are permissible. It is often used with level control for surge tanks as these will often need to maintain a general level rather than a specific value. Furthermore, the controller can be made more aggressive outside the band to quickly drive the system to the desired value.

A more generalised approach to deadbanding is called **gain scheduling**, where the controller gain changes depending on the value of the **scheduling variable**. Most often, the domain of the scheduling variable is partitioned into different regions. For each region, a different controller gain may be assigned. Mathematically, this can be written as

$$u_t = G_c \begin{cases} K_{c,1} & s_t < K_{gs,1} \\ K_{c,2} & K_{gs,1} \leq s_t < K_{gs,2} \\ \vdots & \vdots \\ K_{c,n} & K_{gs,n} < s_t \end{cases} \tag{5.71}$$

where $K_{c,i}$ is the controller gain for the ith region, $K_{gs,i}$ is the ith scheduling limit, and s_t is the scheduling variable. This approach allows for a nonlinear process to be controlled with a series of linear controllers, which would not be possible with a single controller.

5.5.3 Squared Control

Squared control is useful if it is desired to penalise large deviation from the setpoint more strongly than those close to the setpoint. This control law can be written as

$$u_t = K_c \text{sign}(e_t) e_t^2 \tag{5.72}$$

where sign is the sign function that returns −1 if e_t is negative, 0 if the value is zero, and 1 otherwise.

5.5.4 Ratio Control

In certain systems, it may be desirable to keep two variables in a constant ratio, for example, in a mixing process, to keep the composition of the mixture constant, the ratio between the flow rates of the two inlets should be maintained constant. In such cases, we can implement **ratio control** to make sure that the ratio, R, is maintained at the desired value. A schematic for ratio control is shown in Fig. 5.20. The ratio control is implemented by the multiplication block that takes the measured value of the solids and multiplies it by the desired ratio set by the VC controller.

When implementing ratio control, the ratio itself should not be controlled, but rather the ratio controller should set the setpoint for one of the variables using the ratio, for example, if $R = u_1 / u_2$, then we could set $u_2 = Ru_1$ and use this u_2 as the setpoint to the u_2 controller. In all cases, division should be avoided since if one of the variables becomes zero, this will then cause an error. As well, it should be noted that ratio control is implemented using the absolute values of the variables and not their deviational values as is common with other forms of control.

Ratio control is useful when it is desired to control some intensive (amount-invariant) variable, such as composition, density, viscosity, or temperature, using extensive (amount-varying) variables, such as flow rates. Note that when implementing ratio control, we need to be able to scale all extensive variables by the same amount and keep all independent intensive variables constant. Thus, for example, a heat exchanger cannot be effectively controlled by ratio control, since it is not feasible to change the surface area during operation. One solution to the

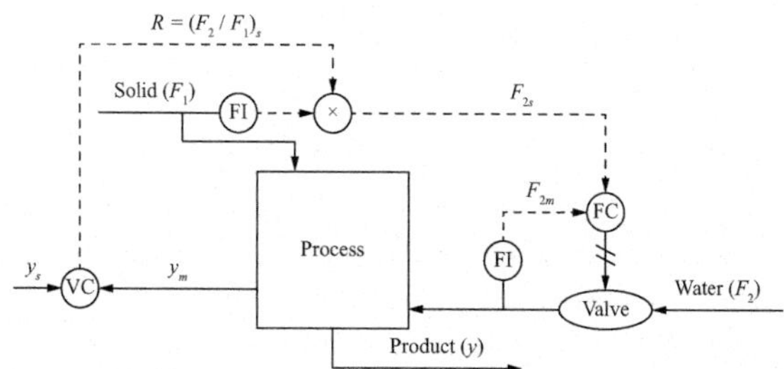

Fig. 5.20 Ratio control with trim feedback control

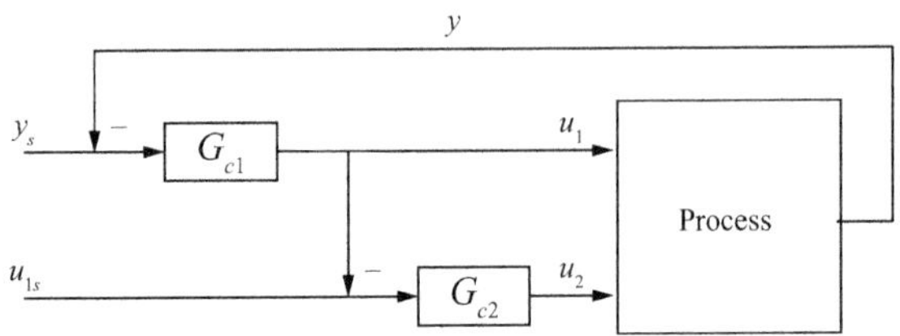

Fig. 5.21 Input-position control

problem of keeping all independent intensive variables constant is to implement a form of feedback control on the ratio itself. Such control is often called **trim feedback control**. In Fig. 5.20, it is shown by the dashed line from the VC to the multiplication block, which determines an appropriate ratio R based on the setpoint and measured values for y. Another advantage of ratio control is that it does not require a model to implement which means that we can obtain feedforward-like control without requiring any models of the system.

5.5.5 Input-Position Control

Input-position control, also called valve-position control, allows controlling multi-input, single-output systems using more advanced methods. In this control strategy, it is normally assumed that there are two inputs: a fast-acting but expensive (or otherwise restricted) input, u_1, and a slow-acting but cheap (or otherwise abundant) input, u_2. An example would be cooling an exothermic reactor using both coolant (expensive, but fast) and cooling water (abundant, but slow). In general, the cheap input is used for controlling the process, but the expensive input can be used to improve the speed of response. Basically, at steady state, it is assumed that the expensive input will attain some steady-state value u_{1s}, so that the overall process will be controlled by u_2. A schematic of this control strategy is shown in Fig. 5.21.

Since input-position control can be treated as a modification of cascade control, the fast loop given by u_1 is tuned first and then the slower loop given by u_2. This also implies that the rule regarding time constants between the loops for cascade control should hold, that is, τ_{c2}/τ_{c1} should be between 4 and 10. In general, both controllers are designed as simple PI controllers without any anti-wind-up features.

5.5.6 Nonlinear Characterisation

Nonlinear characterisation allows nonlinearities present in actuators to be dealt with separately from the controller itself. The nonlinear characterisation block converts the linear controller output into the nonlinear actuator output, for example, if the controller gives a flow rate, then this block can convert the flow rate into the corresponding value. Very often, this block is a look-up table with interpolation between the given data points. Such an approach can allow for the system to be focused on those areas that are highly nonlinear.

5.5.7 *Bumpless Transfer*

Bumpless transfer is the smooth transfer from one control mode or strategy to another without causing any visible, undesirable changes (bumps) in the process variables. This often arises from a mismatch between the requested input values in the two modes. In general, bumpless transfer is ensured using an appropriate anti-wind-up method. However, there is one case, where it may be necessary to take special action and that is when transferring from manual to automatic mode. In these cases, the automatic input value will not be exactly equal to the manual value meaning that it can lead to a bump in the output. The solution is to match the current manual value with the expected automatic value. As well, it may be helpful to reset the integration so that it equals zero at the moment of the switch.

5.6 Further Reading

The following are additional resources on the topics mentioned in this chapter:

(1) **PID Control**:

   a. K. J. Åström and T. Hägglund (1995). *PID controllers: theory, design, and tuning* (Second edition), Research Triangle Park, North Carolina, USA: Instrument Society of America.
   b. A. Visioli (2006). *Practical PID control*, London, UK: Springer-Verlag.

(2) **State Control**: E. D. Sontag (1998). *Mathematical Control Theory: Deterministic Finite Dimensional Systems*, New York, New York, USA: Springer.

(3) **Performance Monitoring**: Y. A. W. Shardt, Y. Zhao, F. Qi, K. H. Lee, X. Yu, B. Huang, and S. Shah (2012). "Determining the State of a Process Control System: Current Trends and Future Challenges" in *The Canadian Journal of Chemical Engineering*, pp. 217–245.

(4) **Feedforward Control**: L. Liu, S. Tian, D. Xue, T. Zhang, and Y. Chen (2019). "Industrial feedforward control technology: a review" in *Journal of Intelligent Manufacturing*, volume 30, pp. 2819–2833.

(5) **Model Predictive Control**: J. B. Rawlings, D. Q. Mayne, and M. M. Diehl (2017). *Model Predictive Control: Theory, Computation, and Design* (Second edition), Madison, United States of America: Nob Hill Publishing.

(6) **Advanced Control Strategies**: S. Skogestad (2023). "Advanced control using decomposition and simple elements" in *Annual Reviews in Control,* volume 56, article 100903.

5.7 Chapter Problems

Problems at the end of the chapter consist of three different types: (a) Basic Concepts (True/False), which seek to test the reader's comprehension of the key concepts in the chapter; (b) Short Exercises, which seek to test the reader's ability to compute the required parameters for a simple data set using simple or no technological aids; and (c) Computational Exercises, which require not only a solid comprehension of the basic material, but also the use of appropriate software.

5.7.1 Basic Concepts

Determine if the following statements are true or false and state why this is the case.

(1) In open-loop control, the controller uses the measured output to determine how to control the process.
(2) In closed-loop control, it is important to be able to measure the output.
(3) Plant-model mismatch can derail open-loop control.
(4) Closed-loop control should provide stable, biasfree, and robust control.
(5) A state-space controller requires measurement of the states.
(6) In state-space control, the separation principle implies that we can separate the design of the observer and the controller.
(7) A proportional-only controller will always provide biasfree control.
(8) The integral term considers the effect of future values on the system.
(9) Integral wind-up occurs when the disturbance changes a lot causing the manipulated variable to fluctuate.
(10) Derivative kick occurs in PID controllers when there is a step change in the setpoint.
(11) Increasing the absolute value of K_c in a PI controller will cause the system to become more stable.
(12) The proportional term in a PID controller considers the current values of the error.
(13) A PI controller can be tuned using the IMC method.
(14) The rise time of a closed-loop system can be used to determine how well a controller regulates the disturbance.
(15) The settling time is defined as three times the closed-loop time constant.
(16) For regulatory-control performance, it is easy to specify the desired values.

(17) Feedforward control seeks to minimise the effects of unmeasurable disturbances on the process.
(18) A decoupler is a type of feedforward control.
(19) In feedforward control, it may be necessary to remove terms such as time delay or unstable zeros from the final controller.
(20) Interlocking occurs when a series of discrete events must be fulfilled before some action can occur.
(21) Supervisory control allows us to create a network of controllers each of which can be controlling another controller.
(22) In cascade control, the primary controller should always be faster than the secondary controller.
(23) Model predictive control allows constraints and economic conditions to be considered in controlling the process.
(24) Model predictive control requires good models.

5.7.2 Short Questions

These questions should be solved using only a simple, nonprogrammable, nongraphical calculator combined with pen and paper.

(25) Describe in words how the following control works. What kind of methods are being used? What are the objectives, and what are some potential disturbances?

a. An elevator reaching a given floor.
b. Temperature control in a home oven.
c. Temperature control in a fridge/freezer.
d. Driving a car on the highway.
e. Driving a car on the highway using cruise control.

(26) Design PI controllers using the formulae in Table 5.1 for the following processes:

a. $G_p = \frac{2}{30s+1}e^{-15s}$
b. $G_p = \frac{-2}{30s+1}e^{-15s}$
c. $G_p = \frac{2}{3s+1}e^{-150s}$
d. $G_p = \frac{-5}{s-0.5}e^{-15s}$

You should use the smallest τ_c possible.

(27) Design PID controllers using the formulae in Table 5.2 for the same processes as in Question (26). You should use the smallest τ_c possible.

5.7.3 Computational Exercises

The following problems should be solved with the help of a computer and appropriate software packages, such as MATLAB®.

(28) Simulate the controllers from Questions (26) and (27) for a setpoint change of +2 and −2. You can assume that there is no disturbance affecting the process. Explain what you see. Which controller would you recommend for each process? Why?

(29) Design dynamic feedforward controllers for the following processes.

a. $G_p = \frac{2}{30s+1}e^{-15s}$, $G_d = \frac{-2}{15s+1}e^{-5s}$

b. $G_p = \frac{-2(5s-1)}{(30s+1)(25s+1)}e^{-15s}$, $G_d = \frac{2}{5s+1}e^{-20s}$

c. $G_p = \frac{2}{(3s+1)(10s+1)}e^{-s}$, $G_d = \frac{-2}{15s+1}e^{-5s}$.

(30) Design static feedforward controllers for the processes in Question (29). Simulate and compare the performance of the two types of feedforward controllers for a setpoint change of +2. You can assume that the disturbance is driven by Gaussian white noise.

(31) Consider a cascade loop with the following two transfer functions:

$$G_{p,\,\text{secondary}} = \frac{2}{3s+1}e^{-s},\; G_{p,\,\text{primary}} = \frac{-2}{30s+1}e^{-20s}. \tag{5.73}$$

Design appropriate PI controllers for this cascade loop. What aspects should you take into consideration when designing this controller? Simulate the system.

(32) Consider the same situation as in Question (31), but now you also wish to ensure that the secondary loop does not have any oscillations and reaches the setpoint as fast as possible. Using simulations, design appropriate controllers for this system.

(33) You need to design a controller for a process whose model has been determined experimentally to be

$$\hat{G}_p = \frac{1}{15s+1}e^{-10s}. \tag{5.74}$$

However, you do not know how well the experimental model reflects the true process. Using the IMC tuning method, design a PID controller assuming that the experimental model is correct. Knowing that the model may be incorrect, do not pick too small a τ_c. Simulate your model assuming Equation (5.74) is correct. If your controller is satisfactory, then try simulating the closed-loop system assuming that the true process model is given by

a. **Small Mismatch**: $G_p = \frac{1.1}{16s+1}e^{-11s}$.
b. **Large Mismatch in Gain**: $G_p = \frac{2}{14s+1}e^{-9s}$.
c. **Large Mismatch in the Time Constant**: $G_p = \frac{1.05}{4s+1}e^{-10s}$
d. **Large Mismatch in Time Delay**: $G_p = \frac{1}{14s+1}e^{-25s}$

If your system is unstable, design a new controller that makes your system stable using both the experimental model and the true model. What conclusions can you draw about controller design?

(34) For the following two-input, two-output system, design a decoupler between u_2 and y_1. The PID couplings are y_1 with u_1 and y_2 with u_2.

$$\begin{bmatrix} y_1 \\ y_2 \end{bmatrix} = \begin{bmatrix} \frac{10}{(10s+1)}e^{-15s} & \frac{-2}{(15s+1)(10s+1)}e^{-10s} \\ \frac{5(10s+1)}{(15s+1)(20s+1)}e^{-25s} & \frac{5}{(10s-1)}e^{-30s} \end{bmatrix} \begin{bmatrix} u_1 \\ u_2 \end{bmatrix} \tag{5.75}$$

(35) Simulate the system given by Eq. (5.75) with and without the decoupler. What is the effect of the decoupler? The PI controllers are given as

$$\begin{aligned} \varepsilon_1 &= \frac{1}{10}\left(1 + \frac{1}{25s}\right)(r_1 - y_1) \\ \varepsilon_2 &= \frac{1}{5}\left(1 + \frac{1}{40s}\right)(r_2 - y_2) \end{aligned} \tag{5.76}$$

(36) Using the DMC method, design a model predictive controller for the following system:

$$y_t = \frac{z^{-10}}{1 - 0.25z^{-1}}u_t \tag{5.77}$$

(37) Using the DMC method, design a model predictive controller for the following system:

$$\begin{bmatrix} y_1 \\ y_2 \end{bmatrix} = \begin{bmatrix} \frac{2}{10s+1}e^{-5s} & \frac{-2}{5s+1}e^{-15s} \\ \frac{5}{15s+1}e^{-10s} & \frac{4}{10s+1}e^{-10s} \end{bmatrix} \begin{bmatrix} u_1 \\ u_2 \end{bmatrix} \tag{5.78}$$

Assume that the sampling time is 1 s.

Chapter 6
Boolean Algebra

Boolean algebra is the algebra of binary variables that can only take two values, for example, `true` and `false` or `1` and `0`. It is very useful for solving problems in logic and is a requirement for good programming. George Boole (1815–1864) is the discoverer of Boolean algebra.

A Boolean expression is a group of elementary terms that are linked with connectors (operators). The mathematical space in which a Boolean algebra is defined will be denoted using a double-struck $\mathbb{B}$.

6.1 Boolean Operators

In Boolean algebra, there are five operators: **conjunction**, **disjunction**, **negation**, **implication**, and **equivalency**. These five operators are described in Table 6.1. It should be noted that negation is a **unary operator**, that is, it only requires a single variable. All other operators are **binary operators**; that is, they require two variables. When we write a Boolean expression, it is common to write conjunction as multiplication and disjunction as addition; that is, $\mathtt{a} \wedge \mathtt{b}$ is written as $\mathtt{ab}$ and $\mathtt{a} \vee \mathtt{b}$ as $\mathtt{a} + \mathtt{b}$.

All Boolean expressions can be written using $\wedge$, $\vee$, $\neg$, 0, and 1. The precedence of the operators is such that $\wedge$ has a higher precedence than $\vee$, for example, $\mathtt{a} \vee \mathtt{b} \wedge \mathtt{c} = \mathtt{a} \vee (\mathtt{b} \wedge \mathtt{c})$.

6.2 Boolean Axioms and Theorems

For the Boolean space $\mathbb{B} = \{0, 1\}$ with variables $\mathtt{a}, \mathtt{b}$, and $\mathtt{c}$ and operators $\wedge$ und $\vee$, the following axioms hold:

Y. A. W. Shardt, *Automation Engineering*,
https://doi.org/10.1007/978-3-031-92394-4_6

Table 6.1 Boolean operators, where $\mathtt{a}, \mathtt{b} \in \mathbb{B}$

Operator	Symbol	Statement	Other representations	Word representation
Conjunction	$\land$	$\mathtt{a} \land \mathtt{b}$	$\mathtt{a} \cdot \mathtt{b}$, $\mathtt{ab}$, a AND b, a & b	AND
Disjunction	$\lor$	$\mathtt{a} \lor \mathtt{b}$	$\mathtt{a} + \mathtt{b}$, a OR b, a||b	OR
Negation	$\neg$	$\neg\mathtt{b}, \mathtt{b}', \overline{\mathtt{b}}$	NOT b, !b	NOT
Implication	$\rightarrow$	$\mathtt{a} \rightarrow \mathtt{b}$	—	IF-THEN
Equivalence	$\leftrightarrow$	$\mathtt{a} \leftrightarrow \mathtt{b}$	—	IF-&ONLY-IF

(1) **Closure**

a. $\mathtt{a} \lor \mathtt{b} \in \mathbb{B}$
b. $\mathtt{a} \land \mathtt{b} \in \mathbb{B}$

(2) **Commutativity**

a. $\mathtt{a} \lor \mathtt{b} = \mathtt{b} \lor \mathtt{a}$
b. $\mathtt{a} \land \mathtt{b} = \mathtt{b} \land \mathtt{a}$

(3) **Distributivity**

a. $\mathtt{a} \land (\mathtt{b} \lor \mathtt{c}) = (\mathtt{a} \land \mathtt{b}) \lor (\mathtt{a} \land \mathtt{c})$
b. $\mathtt{a} \lor (\mathtt{b} \land \mathtt{c}) = (\mathtt{a} \lor \mathtt{b}) \land (\mathtt{a} \lor \mathtt{c})$

(4) **Identity**

a. $\mathtt{a} \lor 0 = \mathtt{a}$
b. $\mathtt{a} \land 1 = \mathtt{a}$

(5) **Complementation**

a. $\mathtt{a} \land \neg\mathtt{a} = 0$
b. $\mathtt{a} \lor \neg\mathtt{a} = 1$

(6) **Annihilator**

a. $\mathtt{a} \lor 1 = 1$
b. $\mathtt{a} \land 0 = 0$

(7) **Associativity**

a. $(\mathtt{a} \land \mathtt{b}) \land \mathtt{c} = \mathtt{a} \land (\mathtt{b} \land \mathtt{c})$
b. $(\mathtt{a} \lor \mathtt{b}) \lor \mathtt{c} = \mathtt{a} \lor (\mathtt{b} \lor \mathtt{c})$

(8) **Idempotence**

a. $\mathtt{a} \lor \mathtt{a} = \mathtt{a}$
b. $\mathtt{a} \land \mathtt{a} = \mathtt{a}$

(9) **Involution**

a. $\neg\neg\mathtt{a} = \mathtt{a}$

(10) **Absorption**

a. $\texttt{a} \wedge (\texttt{a} \vee \texttt{c}) = \texttt{a}$
b. $\texttt{a} \vee (\texttt{a} \wedge \texttt{c}) = \texttt{a}$

With these axioms, it is possible to define a complete Boolean algebra.

An important law is De Morgan's law:

$$\neg(\texttt{a} \vee \texttt{b}) = \neg\texttt{a} \wedge \neg\texttt{b} \tag{6.1}$$

$$\neg(\texttt{a} \wedge \texttt{b}) = \neg\texttt{a} \vee \neg\texttt{b}. \tag{6.2}$$

This law can be used to simplify many complex Boolean expressions or convert between representations.

6.3 Boolean Functions

A Boolean function is a function where all variables are Boolean variables, for example,

$$\texttt{F} = \texttt{f}(\texttt{X}_1, \texttt{X}_2), \text{ where } \texttt{X}_1, \texttt{X}_2 \in \mathbb{B}. \tag{6.3}$$

Typically, a Boolean variable is shown using a capital letter.

A **truth table** shows the values of the expression for each combination of values of the variables, that is, each possible input value. This implies that a truth table will have 2^N rows, where N is the number of input variables, for example, for two variables, we will have $2^2 = 4$ rows and for four variables $2^4 = 16$ rows.

The truth table for the Boolean AND operator is shown in Table 6.2 (left). Since the AND operator requires that both inputs be TRUE, there is only a single row that evaluates to TRUE. For the Boolean OR operator, the truth table is shown in Table 6.2 (right). Since the OR operator requires that at least one of the inputs be TRUE, there are three rows that evaluate to TRUE and one, where both inputs are FALSE that evaluates to FALSE.

Table 6.2 Truth table for the Boolean operators (left) AND and (right) OR

AND			OR		
A	B	AB	A	B	A+B
0	0	0	0	0	0
0	1	0	0	1	1
1	0	0	1	0	1
1	1	1	11	1	1

Example 6.1 Truth Table
What is the truth table for the Boolean function $F = AB'$?

Solution

Since we have two variables (A and B), $N = 2$. This means that there will be $2^2 = 4$ rows in the truth table.

A	B	B′	F
0	0	1	0
0	1	0	0
1	0	1	1
1	1	0	0

There are two common representations for Boolean functions: the **sum-of-products** (SOP) form and the **product-of-sums** (POS) form.

6.3.1 Sum-of-Products Form and Minterms

The sum-of-products form is a representation of a Boolean function where all products (or terms) are products of single variable, for example, $F = ABC + B'CDE' + A'B'$ or $F = BCDE + AB'E + HI' + C$. Using the axioms, it is possible to convert all functions to a sum-of-products form.

Example 6.2 Sum-of-Products Form
Which of the following functions are in the sum-of-product form:

(1) $F = ABC + B'CDE'$
(2) $F = (A + B)C$
(3) $F = ABC + B'(D + E)$
(4) $F = BC + DE'$?

Solution

Only (1) and (4) are in the sum-of-product form, since these representations are a sum of the products of individual variables. In (2) and (3), we have a product of many variables (A and B) in (2) and (D and E) in (3).

Example 6.3 Converting into the Sum-of-Products Form
Convert this function into the sum-of-products form: F = (A + B)C.

Solution

Using the distributive property, we can obtain the sum-of-products form, that is,

$$F = (A + B)C = AC + BC$$

A **minterm** is a row in the truth table where the value is 1. The symbol for a minterm is m_i, where i is the decimal row number. The compact sum-of-products form is $\Sigma m(i,\ldots)$, where i is the row number of the minterms. When the minterms are converted into a functional representation, each variable, whose value is 1, will be written in its plain form, while each variable, whose value is 0 will be written in its negated form, for example, for m_2, the minterm with the binary value 010_2, the term will be given as A′BC′.

Example 6.4 Compact Sum-of-Products Form
What is the compact sum-of-products form for the function F = (A + B)C?

Solution

First, we require a truth table. Then, we can simply find the rows with a value of 1.

A	B	C	F
0	0	0	0
0	0	1	0
0	1	0	0
0	**1**	**1**	**1**
1	0	0	0
1	**0**	**1**	**1**
1	1	0	0
1	**1**	**1**	**1**

The rows with the numbers in bold are the rows where the function has a value of 1. We can find the decimal row number simply by converting the binary representation into a decimal value, that is, for the first bolded row (011).

$$011_2 = 2 + 1 = 3$$

and

$$101_2 = 2^2 + 1 = 5$$

$$111_2 = 2^2 + 2 + 1 = 7$$

Thus, the compact sum-of-products form is

$$\Sigma m(3, 5, 7)$$

It should be noted that the ordering of the Boolean variables is important. When the ordering is changed, then the compact sum-of-products form will also change.

6.3.2 *Product-of-Sums Form and Maxterms*

The product-of-sums form is a representation of a Boolean function, where all the factors are sums of individual variables, for example, $\mathtt{F} = (\mathtt{A} + \mathtt{B} + \mathtt{C})(\mathtt{B}' + \mathtt{C} + \mathtt{D})\mathtt{E}'$ or $\mathtt{F} = (\mathtt{A} + \mathtt{B})\mathtt{C}$. Using the axioms, it is possible to convert all functions into a product-of-sums form.

Example 6.5 Product-of-Sums Form
Which of the following Boolean functions are in the product-of-sums form:

(1) $\mathtt{F} = \mathtt{ABC} + \mathtt{B}'\mathtt{CDE}'$
(2) $\mathtt{F} = (\mathtt{A} + \mathtt{B})\mathtt{C}$
(3) $\mathtt{F} = \mathtt{ABC} + \mathtt{B}'(\mathtt{D} + \mathtt{E})$
(4) $\mathtt{F} = \mathtt{BC} + \mathtt{DE}'$?

Solution

Only (2) and (3) are in the product-of-sums form, since these representations are products of sums with individual variables. In (1) and (4), we have a sum of many variables (A, B, and C) in (1) and (B, C, D, and E) in (4).

Example 6.6 Converting into the Product-of-Sums Form
Convert the function $\mathtt{F} = \mathtt{AC} + \mathtt{BC}$ into its product-of-sums form.

Solution

Using the distributive property, we can obtain the product-of-sums form, that is,

$$F = AC + BC = (A + B)C.$$

A **maxterm** is a row in the truth table with the value of 0. The symbol for a maxterm is M_i, where i is the decimal value of the row. The compact product-of-sums form is $\Pi M(i,\ldots)$, where i is the decimal row value of the corresponding maxterm. When maxterms are converted into a functional representation, then each variable with the value of 0 is converted as written, while each variable with a value of 1 is written in its negated form, for example, M_2, the maxterm with the binary value 010_2, would be written as $A + B' + C$.

Example 6.7 Compact Product-of-Sums Form
What is the compact product-of-sums form for the function $F = (A + B)C$?

Solution

First, we require the truth table for the function. Then, we can easily find the rows with a value of 0.

A	B	C	F
0	**0**	**0**	**0**
0	**0**	**1**	**0**
0	**1**	**0**	**0**
0	1	1	1
1	**0**	**0**	**0**
1	0	1	1
1	**1**	**0**	**0**
1	1	1	1

The rows with bold numbers are the rows where the function has a value of 0. We can obtain the row number by converting the binary values into a decimal representation, for example, for the first row (000).

$$000_2 = 0 + 0 + 0 = 0$$

and

$$001_2 = 0 + 0 + 1 = 1$$

$$010_2 = 0 + 2 + 0 = 2$$

$$100_2 = 2^2 + 0 + 0 = 4$$

$$110_2 = 2^2 + 2 + 0 = 6$$

Therefore, the compact product-of-sums form is

$$\Pi M(0,\ 1,\ 2,\ 4,\ 6)$$

It is important to note that the order of Boolean variables is important. When the order is changed, then the compact product-of-sums form will also change. Also, it is obvious that any terms in the sum-of-products form are not in the product-of-sums form!

6.3.3 Don't-Care Values

It can happen that a given logical situation cannot occur. Such cases are often denoted using $*$ or $\times$ in the truth table. It is then up to the engineer to determine which value (0 or 1) is best. Such cases are called **don't-care values**.

Example 6.8 Don't Cares
What is the truth table for the Boolean function F = AB, when the case A = B = 0 is not possible?

Solution

The truth table is

A	B	F
0	0	×
0	1	0
1	0	0
1	1	1

6.3.4 *Duality*

Duality in Boolean algebra has the following definition:

- Replace AND by OR.
- Replace OR by AND.
- Replace 0 by 1.
- Replace 1 by 0.

Duality is an important property when we are simplifying or minimising an expression. Often the dual of the representation can be easier to work with.

Example 6.9 Dual of a Function
What is the dual of the function $F = (A + B)C$?

Solution

$$F = (A + B)C$$

$$F^D = [(A + B)C]^D$$

$$F^D = [(A + B)]^D + C^D$$

$$F^D = AB + C$$

The dual of a function is often represented by a superscript *D*.

6.4 Minimising a Boolean Function

When a logical expression is to be implemented using electronic gates, it is often important to minimise the expression, that means, that the number of gates required for the implementation is minimised. Often, each Boolean operation requires a separate logic gate. There are three ways to minimise a Boolean function:

(1) **manually** using the axioms and theorems of Boolean algebra;
(2) using a **Karnaugh map**; and
(3) using the **Quine-McCluskey algorithm**, which is often used for large (greater than 10 variables) Boolean functions.

In general, the Karnaugh map is the best method for minimising a Boolean function. A Karnaugh map is a visual algorithm that minimises a Boolean function so that the largest number of groups of `0` and `1` are found. We search for groups of `0` to find the minimal product-of-sums representation and groups for 1 to find the minimal sum-of-products representation. The `0`'s and `1`'s can only be arranged in groups of 2^n, where $n \in \mathbb{N}$. There are simple Karnaugh maps for 2, 3, 4, and 5 variables.

The Karnaugh map for two variables is shown in Fig. 6.1. The diagram is created so that on one axis lies one variable and on the other axis the other variable. Then, for the minimal sum-of-products representation, we circle the largest group with a size that is a power of 2 that covers as many of the `1` as possible.

The Karnaugh map for three variables is shown in Fig. 6.2.The diagram is created so that on one axis lies one variable and on the other axis the remaining two variables. Each row and column has one of the two possible values (`0` or `1`) so that each adjacent row or column only differs by one entry, for example, when the row is `01`, then the next row must be `11`, since only one entry should be changed.

The Karnaugh map for four variables is shown in Fig. 6.3. Each row and column has two variables. The Karnaugh map for five variables is shown in Fig. 6.4. When we are looking for `1`, it is common to not bother with writing the `0`.

The general procedure for finding the minimal sum-of-products form is:

(1) Construct the Karnaugh map.
(2) Circle the `1` into the largest groups possible.
(3) Write the corresponding sum-of-product form based on the circled groups. For each group, we only write the variables whose value is not changed. Variables with a value of 0 are negated, for example, in Fig. 6.3, for the green loop, `C` varies between `0` and `1`, which means that it is ignored. The other variables remain constant. `A` and `D` have a value of `1`, while `B` has a value of `0`. Thus, we negate `B` and write the other two variables normally to give `AB'D`.

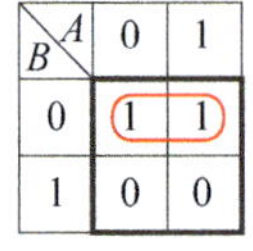

Fig. 6.1 Karnaugh map for the function `F = B'`

A	B \ C	0	1
0	0	1	0
0	1	0	1
1	1	0	0
1	0	0	1

Fig. 6.2 Karnaugh map for the function `F = Σm(0, 3, 5)`

C D \ A B	0 0	1 0	1 1	0 1
0 0	1	0	0	1
0 1	0	1	0	0
1 1	1	1	1	1
1 0	1	1	1	1

Fig. 6.3 Karnaugh map for the function `F = AB'D + A'D' + C`

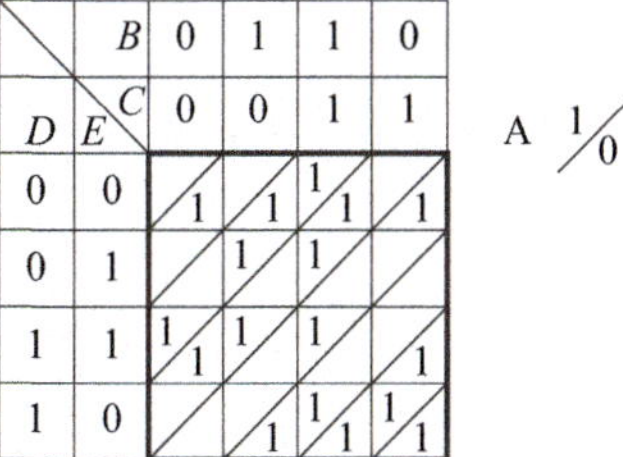

Fig. 6.4 Karnaugh map for the function F = ΠM(`1, 2, 5, 9, 11, 13, 15, 16, 17, 18, 20, 21, 23, 24, 26`)

The general procedure for finding the minimal product-of-sums form is:

(1) Construct the Karnaugh map.

(2) Circle the `0` into the largest groups possible.

(3) Write the corresponding sum-of-product form for `F'` based on the circled groups. For each group, we only write the variables whose value is not changed. Variables with a value of `0` are negated, for example, in Fig. 6.3, for the loop in the first row, `A` varies between `0` and `1`, which means that it is ignored. The other variables remain constant and have a value of `0`. Thus, we write the negated forms of these variables, that is, `B'C'D'`.

(4) Convert `F'` to `F`.

Of course, we can look for the loops using trial and error, but we have no guarantee that we will find the best loops. Therefore, we wish to find a procedure that we can always find the best loops.

Before we can describe such a procedure, we need to define certain words. An **implicant** is a single `1` or group of `1`'s. A **prime implicant** is an implicant that cannot be further combined, that means, that a single `1` creates a prime implicant if there are no adjacent `1`'s. Two adjacent `1`'s create a prime implicant if they cannot be combined into a group of four `1`'s, while four adjacent `1`'s create a prime implicant if they cannot be combined into a group of eight `1`'s. Figure 6.5 shows the difference between an implicant and a prime implicant. We can state that a sum-of-products form with nonprime implicants is not a minimal form, but not all prime implicants will necessarily be required for the minimal sum-of-products form. An **essential prime implicant** is a minterm that is covered by only a single prime implicant. All essential prime implicants must be in the minimal sum-of-products form. Thus, we can say that the objective is to find all the essential prime implicants. Figure 6.7 shows the procedure for minimising a Karnaugh map.

Example 6.10 Karnaugh map

For the Karnaugh map shown in Fig. 6.6, find the minimal sum-of-products form.

(continued on the bottom of the next page.)

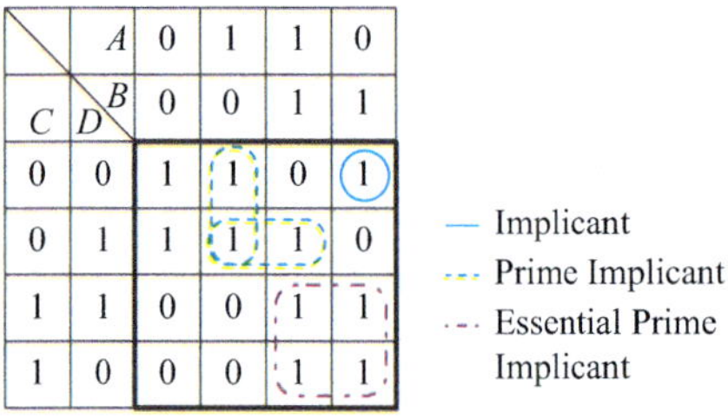

	A	0	1	1	0
C	*D* \ *B*	0	0	1	1
0	0	1	1	0	1
0	1	1	1	1	0
1	1	0	0	1	1
1	0	0	0	1	1

— Implicant
--- Prime Implicant
-·- Essential Prime Implicant

Fig. 6.5 Implicant, prime implicant, and essential prime implicant

	A	0	1	1	0
C	*D* \ *B*	0	0	1	1
0	0	1	1	0	0
0	1	1	1	0	0
1	1	0	1	1	1
1	0	1	0	0	0

Fig. 6.6 Karnaugh map for Example 6.10

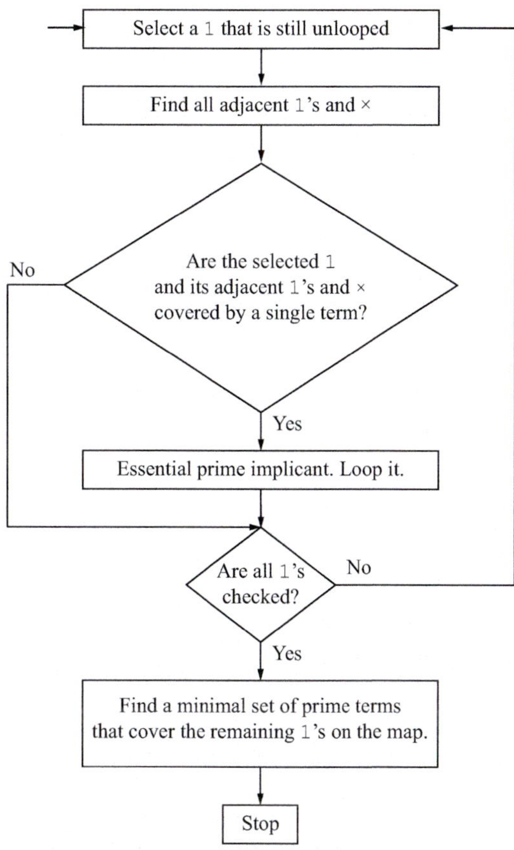

Fig. 6.7 Procedure for minimising a Karnaugh map

Solution

(1) We start with the first `1` (cell: 0000). The largest group that we can find is four `1`. This group is an essential prime implicant.

	A	0	1	1	0
C	D \ B	0	0	1	1
0	0	1	1	0	0
0	1	1	1	0	0
1	1	0	1	1	1
1	0	1	0	0	0

(2) The next unlooped 1 is in the cell 0010. This 1 can only be combined with the 1 in cell 0000. This group is also an essential prime implicant.

	A	0	1	1	0
C	D \ B	0	0	1	1
0	0	1	1	0	0
0	1	1	1	0	0
1	1	0	1	1	1
1	0	1	0	0	0

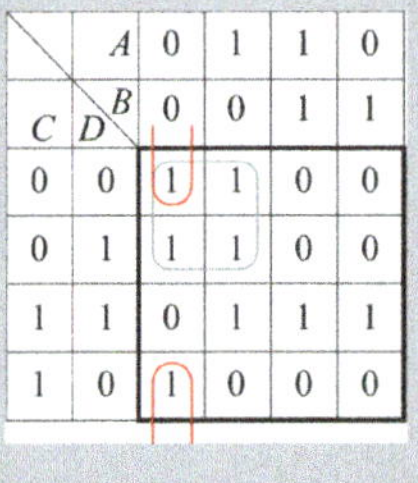

(3) The next free 1 is in cell 0111. Here we have two possibilities. We could make the loop with either cell 0101 or cell 1111. This implies that we have a prime implicant.

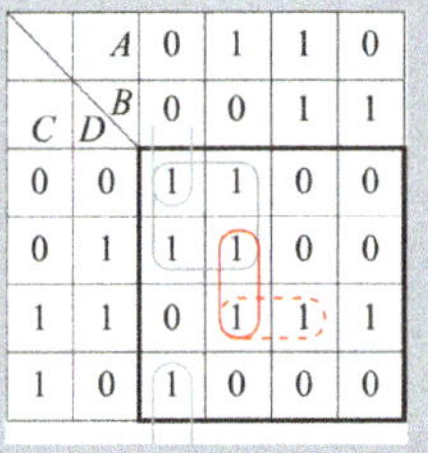

	A	0	1	1	0
C	D \ B	0	0	1	1
0	0	1	1	0	0
0	1	1	1	0	0
1	1	0	1	1	1
1	0	1	0	0	0

(4) The next free 1 is in cell 1011, which we can combine with cell 1111. This is the only possibility for combining this cell. Therefore, this group is also an essential prime implicant.

	A	0	1	1	0
C	D \ B	0	0	1	1
0	0	1	1	0	0
0	1	1	1	0	0
1	1	0	1	1	1
1	0	1	0	0	0

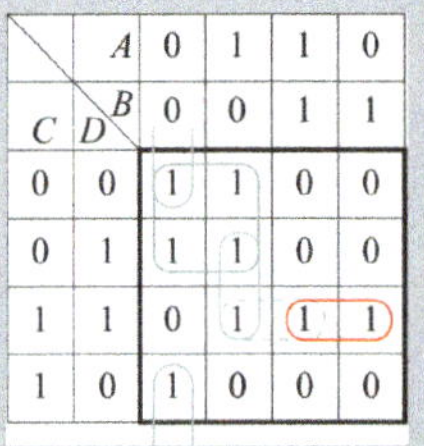

(5) The last step is to write the minimal sum-of-product form. First, we will write all the essential prime implicants, that is, $A'B'D'$, $B'C'$, and BCD Then, we must select one of the remaining prime implicants ($AB'D$ or ACD). There is no difference which of these two prime implicants we select. The minimal sum-of-products form is then

$$F = A'B'D' + B'C' + BCD + \{AB'D \text{ or } ACD\}$$

6.5 Further Reading

The following are additional resources on the topics mentioned in this chapter:

(1) P. Halmos and S. Givant (2008). *Introduction to Boolean Algebras*, New York, New York, USA: Springer.

6.6 Chapter Problems

Problems at the end of the chapter consist of two different types: (a) Basic Concepts (True/False), which seek to test the reader's comprehension of the key concepts in the chapter; and (b) Short Exercises, which seek to test the reader's ability to compute the required parameters for a simple data set using simple or no technological aids.

6.6.1 *Basic Concepts*

Determine if the following statements are true or false and state why this is the case.

(1) In Boolean algebra, there are three possibilities: – 1, 0, and 1.
(2) In Boolean algebra, conjunction is denoted using $\wedge$.
(3) In Boolean algebra, negation is denoted using $\vee$.
(4) The Boolean statements $\neg\mathtt{b}$ and $\mathtt{b}'$ have the same meaning.
(5) The Boolean operator $\wedge$ is a unary operator.
(6) If $\mathtt{a} = \mathtt{1}$, $\mathtt{b} = \mathtt{0}$, and $\mathtt{c} = \mathtt{1}$, then the Boolean statement $\mathtt{a} \vee \mathtt{b} \wedge \mathtt{c}$ evaluates to false.
(7) Boolean algebra does not obey the law of associativity.
(8) Boolean algebra obeys the law of distributivity.
(9) De Morgan's law states that $(\mathtt{A} + \mathtt{B})' = \mathtt{A}'\mathtt{B}'$.
(10) The function $\mathtt{F} = (\mathtt{A} + \mathtt{B})\mathtt{C}$ is in a sum-of-products form.
(11) The function $\mathtt{F} = \mathtt{ABC} + \mathtt{A}'\mathtt{B}'\mathtt{C}'$ is in a sum-of-products form.
(12) The function $\mathtt{F} = (\mathtt{A} + \mathtt{B})\mathtt{C}$ is in a product-of-sums form.
(13) The function $\mathtt{F} = \mathtt{ABC} + \mathtt{A}'\mathtt{B}'\mathtt{C}'$ is in a product-of-sums form.
(14) For a function with the representation $\mathtt{F} = \Sigma m(\mathtt{1}, \mathtt{3}, \mathtt{5})$, the minterm m_2 is equal to $\mathtt{1}$.
(15) For the function with the representation $\mathtt{F} = \Pi M(\mathtt{1},\ \mathtt{3},\ \mathtt{5})$, the maxterm M_3 is equal to 0.
(16) The dual of $\mathtt{F} = \mathtt{A} + \mathtt{BC}$ is $\mathtt{F}^{\mathtt{D}} = \mathtt{A}(\mathtt{B} + \mathtt{C})$.
(17) We must always assign a don't-care a value of $\mathtt{1}$.
(18) All prime implicants are in the minimal sum-of-products form.

(19) All essential prime implicants are in the minimal sum-of-products form.
(20) A Karnaugh map can be used to determine the minimal product-of-sums form.

6.6.2 Short Questions

These questions should be solved using only a simple, nonprogrammable, non-graphical calculator combined with pen and paper.

(21) Simplify the following function $Z = (A + B)(A + BC)(B + BC)$. Using a truth table, show that the original and simplified functions are the same.
(22) Convert the following functions into the sum-of-products form:

a. $Z = (A + B)(A + C)(A + D)(BDC + E)$
b. $Z = (A + B + C)(B + C + D)(A + C)$

(23) Convert the following functions into a product-of-sums form:

a. $Z = W + XYZ$
b. $Z = ABC + ADE + ABE$

(24) Find the compact product-of-sums form for the following functions:

a. $F(A, B, C) = A'$
b. $F(A, B, C, D) = A'B' + A'B'(CD + CD')$

(25) Find the compact sum-of-products form for the following functions:

a. $F(A, B, C) = (A' + B + C)(A + C)$
b. $F(A, B, C, D) = A'B' + A'B'(CD + CD')$

(26) Using a Karnaugh map, find the minimal sum-of-products form for the following functions:

a. $F(A, B, C, D) = A'B' + A'B'(CD + CD')$
b. $F(A, B, C, D) = \Sigma m(0, 1, 2, 3, 6, 7)$
c. $F(A, B, C, D) = \Sigma m(0, 4, 5, 6, 8, 9, 10, 11)$
d. $F(A, B, C, D) = \Sigma m(10, 12, 14)$

(27) For the Karnaugh map shown in Fig. 6.8, find all prime implicants and essential prime implicants.

Fig. 6.8 Karnaugh map for Question 27

	A	0	1	1	0
C	D \ B	0	0	1	1
0	0	1	0	0	0
0	1	1	1	0	0
1	1	0	1	0	0
1	0	0	1	1	0

Chapter 7
PLC Programming

When using a PLC, it is necessary to transfer the required information to the PLC using some methods. There are two possible approaches. We could use an *ad hoc* approach that depends on the PLC or situation, or we could use some type of standard. It is obvious that using standards is better since we can re-use the information, for example, the same code can be re-used for different cases without having to worry about compatibility problems. The IEC/EC 61131 is the standard for PLC programming and will be described in the following sections. This standard consists of five different programming languages and a common set of rules that apply to all the programming languages.

7.1 The Common IEC-Standard Hierarchy

The foundational component of the IEC 61131-3 standard is the **programme organisation unit** (POU). The POU is the smallest self-standing component of a PLC programme. There are three types of POUs:

(1) **Function (`FUN`)**: A function is a parametrisable POU without any static variables or state information that given the same input parameters will give the same output.
(2) **Function Block (`FB`)**: A function block is a parametrisable POU with static variables (that is, *memory*). Given the same input values, a function block can give different values that depend on the internal function-block values, as well as external values.
(3) **Programme (`PROG`)**: A programme represents the *main programme*. All the variables for the programme and their physical addresses must be specified. Otherwise, a programme is like a function block.

Y. A. W. Shardt, *Automation Engineering*, https://doi.org/10.1007/978-3-031-92394-4_7

Programmes and function blocks can have input and output parameters, while functions only have input parameters and the function value as the return value. A POU consists of three parts:

(1) Declaration of the POU type with POU name (and data type for functions). The possibilities are:

   a. **Function**: `FUNCTION` *`Name DataType`* ... `END_FUNCTION`
   b. **Function Block**: `FUNCTION_BLOCK Name ... END_FUNCTION_BLOCK`
   c. **Programme**: `PROGRAM Name ... END_PROGRAM`

(2) Declaration of variables
(3) Remainder of the POU with the instructions.

Before a POU can be used, each programme must be associated with a task. Before we can load a task into the PLC, we must first define:

(1) Which **resources** do we need? In the standard, a resource is defined as either a central processing unit (CPU) or a special processor.
(2) How is the programme executed and with what priority?
(3) Do variables have to be assigned to physical PLC addresses?
(4) Do references to other programs have to be made using global or external variables?

The priority of a task shows how the given task should be executed. There two important parts to define the priority: **scheduling** and **priority type**. Scheduling focuses on how the programme is executed. There are two possibilities: **cyclically**, that is, the programme will be continually executed; and **on demand**, that is, the programme will only be executed as needed. The priority type represents whether the programme can be interrupted. Again, there are two possibilities: **non-pre-emptive** and **pre-emptive**. A non-pre-emptive task must always be completed before another task can be started. A pre-emptive task can be stopped if a task of higher priority occurs. Figure 7.1 shows the two possibilities. As well, the priority level needs to be set. This ranges from 0 for the highest priority to 3 for the lowest priority.

Defining these components creates a **configuration**. Figure 7.2 shows a visual representation of the components of a configuration and how these are combined together.

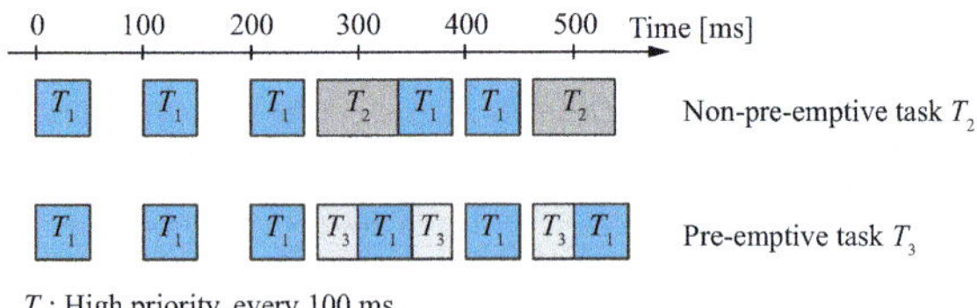

Fig. 7.1 Pre-emptive and non-pre-emptive tasks

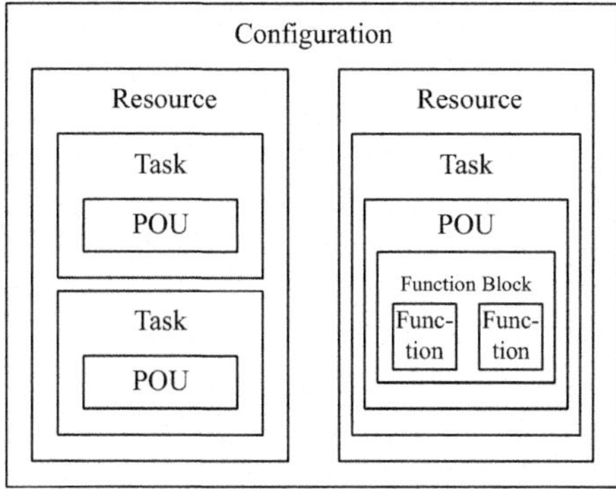

Fig. 7.2 Visual representation of a configuration

The call hierarchy is defined as follows: a programme can only call function blocks and functions. A function block can only call other function blocks and functions, while a function can only call other functions. Recursion cannot be implemented in this standard. POUs cannot call themselves nor can they be called as a result of a chain of POUs.

7.2 Types of Variables

In the IEC standard, there are many different types of variables:

(1) **Variables** (`VAR`): These are general variables that can be used by all POUs.

(2) **Input Variables** (`VAR_INPUT`): The actual parameter will be transferred by value to the POU, that is, the variable itself is not passed to the POU, but only a copy. This ensures that the input variable outside of the POU cannot be changed. This concept is also often called **call by value**. All POUs can use this type.

(3) **Output Variables** (`VAR_OUTPUT`): The output variable will be returned to the calling POU as a value. This concept is also called **return by value**. All POUs can use this type.

(4) **Input and Output Variables** (`VAR_IN_OUT`): The actual parameter will be transferred to the calling POU as a **pointer**, that is, the location of the variable will be given so that any changes made to the variable are directly stored. The concept is also called **call by reference**. For variables with complex data structures, this can lead to efficient programming. However, since the variable's location is passed, this means that (undesired) changes will affect the variable even outside the calling function. All POUs can use this type.

(5) **External Variables** (`VAR_EXTERNAL`): This variable can be changed outside of the POU using the variable. An external variable is required to have read and write access to a global variable of another POU within a given POU. It is only visible for POUs that list this global variable under `VAR_EXTERNAL`; all others have no access to this global variable. The identifier and type of a variable

under VAR_EXTERNAL must match the corresponding VAR_GLOBAL declaration in the program. Only programmes and function blocks can use this type.

(6) **Global Variables** (VAR_GLOBAL): A globally declared variable can be read by and written by several POUs. To do this, the variable must be specified in the other POUs under the VAR_EXTERNAL with the same name and type.

(7) **Access Variables** (VAR_ACCESS): Access variables are global variables for the configurations that act as a communication channel between components (resources) of the configurations. It can be used like a global variable within the POU.

7.3 Variables, Data Types, and Other Common Elements

The IEC standard defines common elements that apply to all programmes. These elements are not only components, but also rules that determine how the elements can be used.

7.3.1 Simple Elements

Each PLC programme consists of basic elements that are defined as the smallest unit that, when combined together, build declarations and instructions, which form a complete programme. These simple elements can be classified into **delimiters**, **keywords**, **literals**, and **identifiers**.

Delimiters

Delimiters are symbols that separate the individual components from one another. Typical delimiters include the space, +, the comma (,), and *. Table 7.1 shows all the delimiters in the IEC standard.

Keywords

In the IEC 61131-3 standard, keywords are the elementary *word*. Normally, keywords are written in **bold**. They are standard identifiers that are clearly specified in the IEC 61131-3 standard in terms of spelling and purpose. Therefore, they cannot be used in a user-defined manner for variable or other names. Capitalisation is not

Table 7.1 Delimiters in the IEC 61131-3 standard

Delimiter	Meaning, Clarification
Space	Can be inserted anywhere, except within keywords, literals, identifiers, directly represented variables, or combinations of delimiters, such as `(*` or `*)`. IEC 61131-3 does not specifically make any statements about tabs, and hence, they are usually treated as spaces
End-of-Line	At the end of an instruction line in instruction list, in structured text also permitted within an instruction. Not permitted within comments in instruction list
Start of Comment `(*`	Starts a comment (not nestable)
End of Comment `*)`	Ends a comment
Plus +	1. Leading sign of a decimal literal 2. In the exponent of a floating-point literal 3. Addition operator in expressions
Minus -	1. Leading sign of a decimal literal 2. In the exponent of a floating-point literal 3. Negation operator in expressions 4. Year-month-day separator in time literals
Octothorpe #	1. Base-number separator in literals 2. Time-literal separator
Period `.`	1. Integer/fraction separator 2. Separator within hierarchical addresses of directly represented and symbolic variables 3. Separator between components of a data structure (when accessing it) 4. Separator for components of a `FB` instance (when accessing it)
e, E	Leading character for exponents of floating-point literals
Quotation Mark `'`	Start and end of strings
Dollar Sign `$`	Start of a special symbol in a string

Prefix for Time Literal
`t#,T#; d#, D#;`
`d, D; h, H; m, M;`
`s, S; ms, MS;`
`date#, DATE#;`
`time#, TIME#;`
`time_of_day#;`
`TIME_OF_DAY#;`
`tod#, TOD#;`
`date_and_time#;`
`DATE_AND_TIME#;`
`dt#, DT#`
	Introductory characters for time literals, combinations of lowercase and uppercase letters are permitted

(continued)

Table 7.1 (continued)

Delimiter	Meaning, Clarification
Colon :	Separator for: 1. Time within time literals 2. Definition of the data type for variable declaration 3. Definition of a data-type name 4. Step names 5. `PROGRAM`... WITH... 6. Function name/data type 7. Access path: data/type 8. Jump label before the next instruction 9. Network name before the next instruction
Assignment (Walrus) Operator :=	1. Operator for initial-value assignment 2. Input connection operator (assignment of actual parameters to formal parameters when calling the POU) 3. Assignment operator
(Round) Brackets (...)	Start and end of: 1. Initial-value list, also: multiple initial values (with repetition number) 2. Range specification 3. Field index 4. Sequence length 5. Operator in instruction list (calculation level) 6. Parameter list when calling the POU 7. Subexpression hierarchy
Square Brackets [...]	Start and end of: 1. Array index (access to an array) 2. String length (when declaring a string)
Comma ,	Separator for: 1. Lists 2. Initial-value lists 3. Array indices 4. Variable names (when there are multiple variables with the same data type) 5. Parameter list when calling a POU 6. Operator in instruction list 7. `CASE` list
Semicolon ;	End of: 1. Definition of a (data) type 2. Declaration (of a variable) 3. Structured-text command
Period-Period ..	Separator for: 1. Range specifications 2. `CASE` branches
Percent %	Introductory character for hierarchical addresses for directly represented and symbolic variables

(continued)

Table 7.1 (continued)

Delimiter	Meaning, Clarification
Assignment Operator `=>`	Output binding operator (assignment of formal parameters to actual parameter when calling a `PROGRAM` or `FUNCTIONBLOCK`)
Comparison `>, <; >=, <=; =, <>`	Comparison operators in expressions
Exponent `**`	Exponentiation in expressions
Multiplication `*`	Multiplication in expressions
Division `/`	Division in expressions
Ampersand `&`	AND operator in expressions

significant for keywords, *i.e.* they can be written using all lowercase, all uppercase, or a mixture of the two. In this book, all keywords are shown in capitals. The keywords include the following possibilities:

(1) names of elementary data types;
(2) names of standard functions;
(3) names of standard function blocks;
(4) names of the input variables for the standard functions;
(5) names of the input and output variables of the standard function blocks;
(6) the variables `EN` and `ENO` in the graphical programming languages;
(7) the operators of the instruction-list language;
(8) the elements of the structured-text language; and
(9) the elements of sequential charts.

Table 7.2 shows all the keywords in the IEC standard (Table 7.3).

Literals

Literals are used to represent the value of a variable (constant). They depend on the data types of these variables. A distinction is made between the following three basic types:

(1) **Numerical Literals** that give the numeric value of a number as a bit sequence, as well as integer and floating-point numbers.
(2) **String Literals** that give the value of a string in either single- or double-byte representation. A string literal is delimited by single, straight quotation marks (`'`; U+0027), for example, `''` is the empty string literal and `'Automation Engineering!'`. If we wish to use a reserved symbol in a string literal, we must place a dollar sign before the reserved symbol, for example, `'$$45'` will give “\$45”. As well, there are various nonprintable special characters that can be represented using the dollar sign. Table 7.3 lists some of the more common special characters.
(3) **Time Literals** that give the value for time points, durations, and dates.

Table 7.2 All keywords in the IEC standard

A		
ABS	ANY	ANY_MAGNITUDE
ACOS	ANY_BIT	ANY_NUM
ACTION	ANY_DATE	ANY_REAL ARRAY
ADD	ANY_DERIVED	ASIN
AND	ANY_ELEMENTARY	AT
ANDN	ANY_INT	ATAN
B		
BOOL	BY	BYTE
C		
CAL	CLK	CTU
CALC	CONCAT	CTUD
CALCN	CONFIGURATION	CU
CASE	CONSTANT	CV
CD	COS	
CDT	CTD	
D		
D	DINT	DT
DATE	DIV	DWORD
DATE_AND_TIME	DO	
DELETE	DS	
E		
ELSE	END_PROGRAM	EN
ESIF	END_REPEAT	ENO
END_ACTION	END_RESOURCE	EQ
END_CASE	END_STEP	ET
END_CONFIGURATION	END_STRUCT	EXIT
END_FOR	END_TRANSITION	EXP
END_FUNCTION	END_TYPE	EXPT
END_FUNCTION_BLOCK	END_VAR	
END_IF	END_WHILE	
F		
FALSE	FIND	FUNCTION
F_EDGE	FOR	FUNCTION_BLOCK
F_TRIG	FROM	
G		
GE	GT	
I		
IF	INITIAL_STEP	INT

(continued)

Table 7.2 (continued)

IN	INSERT	INTERVAL
J		
JMP	JMPC	JMPCN
L		
L	LEN	LREAL
LD	LIMIT	LT
LDN	LINT	LWORD
LE	LN	
LEFT	LOG	
M		
MAX	MOD	MUX
MID	MOVE	
MIN	MUL	
N		
N	NEG	NOT
NE	NON_RETAIN	
O		
OF	OR	
ON	ORN	
P		
P	PROGRAM	PV
PRIORITY	PT	
Q		
Q	QU	
Q1	QD	
R		
R	RELEASE	RETCN
R1	REPEAT	RETURN
R_EDGE	REPLACE	RIGHT
R_TRIG	RESOURCE	ROL
READ_ONLY	RET	ROR
READ_WRITE	RETAIN	RS
REAL	RETC	
S		
S	SIN	STEP
S1	SINGLE	STN
SD	SINT	STRING
SEL	SL	STRUCT

(continued)

Table 7.2 (continued)

SEMA	SQRT	SUB
SHL	SR	
SHR	ST	
T		
T	THEN	TO
TAN	TIME	TOD
TASK	TIME_OF_DAY	TOF
TON	TRANSITION	TYPE
TP	TRUE	
U		
UDINT	ULINT	USINT
UINT	UNTIL	
V		
VAR	VAR_EXTERNAL	VAR_IN_OUT
VAR_ACCESS	VAR_GLOBAL	VAR_OUTPUT
VAR_CONFIG	VAR_INPUT	VAR_TEMP
W		
WHILE	WORD	
WITH	WSTRING	
X		
XOR	XORN	

Table 7.3 Special strings

Dollar-sign representation	Representation on the screen or printer
`$nn`	Shows "nn" in hexadecimal in ASCII
`$$`	$
`$', $"`	', "
`$L, $l`	Line feed (`$0A`)
`$N, $n`	New line
`$P, $p`	New page
`$R, $r`	Carriage return[1] (`$0D`)
`$T, $t`	Tab

The octothorpe is used to give additional information about the literal, for example, `2#` implies a binary representation. The additional information always comes before the octothorpe. Common representations are:

[1]This term comes from the days of typewriters, where the writing head with the ink ribbon was attached to a carriage that had to be pushed at the end of each line to the beginning of the new line.

(1) **Binary Representation**: `2#`.
(2) **Hexadecimal Representation**: `16#`.
(3) **Duration Representation**: `T#` or `TIME#`.
(4) **Date Representation**: `D#` or `DATE#`.
(5) **Time-of-Day Representation**: `TOD#` or `TIME_OF_DAY#`.
(6) **Date-and-Time Representation**: `DT#` or `DATE_AND_TIME#`.

It is possible to use any defined data structure with the octothorpe. Numerical and time literals may also contain underscores in order to make the presentation more legible. Capitalisation is not important.

Time literals have some special properties. There are different types of time literals: duration, date, time of day, and date and time. For each case, there is a special representation with its own rules.

Duration is represented by `T#`. After the octothorpe, the duration is given using the following units:

(1) `d`: Day;
(2) `h`: Hours;
(3) `m`: Minutes;
(4) `s`: Seconds;
(5) `ms`: Milliseconds.

Each unit is separated by an underscore.[2] The units must be placed from largest to smallest. The smallest unit can have a decimal value, for example, `T#1m_10s_100.7ms`. The highest value can "overflow", for example, the time duration `T#127m_19s` is valid and will be automatically converted into the proper representation of `T#2h_7m_19s`. A negative value is also possible, for example, `T#-22s_150ms`.

Dates are represented using `D#`. After the octothorpe comes the date in scientific notation, that is, year-month-day, for example, `D#2017-02-28`.

The time of day is represented using `TOD#`. After the octothorpe comes the time in the format `Hours:Minutes:Seconds.Decimal Part`, for example, `TOD#12:45:25.21`. Note that the 24-h clock is used.

The date and time are shown using `DT#`. After the octothorpe comes the date in scientific notation followed by a dash and the time in the time-of-day format, for example, `DT#2017-05-30-2:30:12`.

Identifiers

Identifiers are alphanumeric strings that allow the PLC programmer to give individual names to the variables, programmes, and related elements. These include jump and network names, configurations, resources, tasks, runtime programmes,

---

[2] The underscore is not obligatory to separate the units, but it does help with the readability of the text.

functions, function blocks, access paths, variables, derived data types, structures, transitions, steps, and action blocks. The identifiers must satisfy the following rules:

(1) The first element cannot be a number (✗ `1Prog`).
(2) No more than one underscore can be used together (✗ `A__B` [with two underscores]).
(3) No delimiters can be used (✗ `w34$23`).

Only the first six characters are considered when comparing two identifiers; that is, both `TUI_123` and `TUI_125` are equivalent. Capitalisation also plays no role, that is, `TUI`, `tui`, and `TuI` are all equivalent.

7.3.2 Variables

A variable is a representation of a physical memory location on a PLC to which a meaning has been assigned. The variable declaration block is bracketed with `VAR_`*`type`* and `VAR_END`. It is possible to specify the type of variable. Each variable is declared on its own line with the following key components:

```
Variable_name: Data_type:= Initial_value;
```

The components in bold must always be given, while the values of components in italics are specified by the programmer. The initial value needs not be given, but it is always better to do so.

7.3.3 Data Types

There are two data types: elementary and derived data types.

Elementary Data Type

An elementary data type is defined as a simple data type that is predefined in the IEC standard. The elementary data types are characterised by their data size (number of bits) and the data range. Both values are defined by the standard, except for dates, times, and strings, whose data size and range depend on the implementation. Table 7.4 shows the elementary data types with their properties (data range and default initial value). The IEC 61131-3 standard defines five groups of elementary data types that can be referenced by the given general data type given in brackets:

(1) **Bit Sequence and Boolean** (ANY_BIT)
(2) **Signed and Unsigned Integers** (ANY_INT)
(3) **Floating-Point Numbers** (ANY_REAL)
(4) **Dates and Times** (ANY_DATE)
(5) **Strings and Durations** (ANYSTRING, TIME).

The general data types ANY_INT and ANY_REAL can be represented together by the group name ANY_NUM.

For an elementary data type, it is possible to define the initial value and range. The initial value is defined as the value of the variable when it is first used. The range defines what the possible values for the variable are.

Table 7.4 Elementary data types in the IEC 61131-3 standard

Data type	Keyword	Bits	Range	Initial value
BOOL	Boolean	1	{0, 1}	0
BYTE	Bit sequence 8	8	[0, 16#FF]	0
WORD	Bit sequence 16	16	[0, 16#FFFF]	0
DWORD	Bit sequence 32	32	[0, 16#FFFF FFFF]	0
LWORD	Bit sequence 64	64	[0, 16#FFFF FFFF FFFF FFFF]	0
SINT	Short Integer	8	$[-128, +127]$	0
INT	Integer	16	$[-32\,768, +32\,767]$	0
DINT	Double Integer	32	$[-2^{31}, +2^{31}-1]$	0
LINT	Long Integer	64	$[-2^{63}, +2^{63}-1]$	0
USINT	Short Integer	8	$[0, +255]$	0
UINT	Integer	16	$[0, +65\,535]$	0
UDINT	Double Integer	32	$[0, +2^{32}-1]$	0
ULINT	Long Integer	64	$[0, +2^{64}-1]$	0
REAL	Floating-Point	8	*see IEC 60559*	0
LREAL	Long Floating-Point	16	*see IEC 60559*	0
DATE	Date	—	—	d#0001-01-01
TOD	Time of Day	—	—	tod#00:00:00
DT	Date with Time of Day			dt#0001-01-01-00:00:00
TIME	Duration	—	—	t#0s
STRING	(Single) String	—	—	''
WSTRING	Double (String)	—	—	""

The initial letters in the data types represent: D = double, L = long, S = short, and U = unsigned

Arrays

Arrays are data elements of identical type that are sequentially stored in memory. An array element can be accessed with the help of an array index that lies within the array boundaries. The value of the index indicates which array element is to be accessed. Most PLC systems ensure that array access with an array index outside the array limits results in an error message during runtime. The array is defined using square brackets ([]). The dimensions are separated by commas, *e.g.*

```
ARRAY [1...45] OF INT
```

is a one-dimensional array with 45 elements of data type `INT`, while

```
ARRAY [1...50,1...200] OF INT.
```

is a two-dimensional array of data type `INT` with 50 elements in one dimension and 200 elements in the other dimension.

The elements of an array are accessed using square brackets, for example, `TEST[3]` takes the third element in the array `TEST`. The dimensions are also separated using commas.

The initial values in the array can be defined using square brackets. When values are repeated, then we can use the format `Repeats(Values)`, to simplify matters, for example, `2(4)` means that we will write the value `4` twice. Thus, the following two array definitions are equivalent:

```
TEST1: ARRAY [1...5] OF INT:= [1, 1, 1, 3, 3];
TEST1: ARRAY[1...5] OF INT:= [3(1), 2(3)];
```

Data Structures

Using the keywords `STRUCT` and `END_STRUCT`, it is possible to define new hierarchical data structures that contain arbitrary elementary or other already defined derived data types as subelements. If a subelement is in turn a structure, a hierarchy of structures is created, for which the lowest structure level is formed from elementary or derived data types.

As in many other programming languages, the components of a data structure are accessed using a period and the component name, for example, `VAR.TEST[3]`, where the component `TEST` is an array.

Derived Data Types

A **derived data type** is a user-defined data type that consists of elementary data types, arrays, and data structures. This procedure is called **derivation** or **type definition**. In this way, a programmer can define the best data model for the problem at hand. A derived data type is defined by `TYPE` and `END_TYPE`. The initial values for the elements of a derived data type are given by :=, for example,

```
TYPE
   COLOUR : (red, yellow, green);
   SENSOR : INT;
   MOTOR :
         STRUCT
              REVOLUTIONS : INT := 0;
              LEVEL : REAL := 0;
              MAX : BOOL := FALSE;
              FAILURE : BOOL := FALSE;
              BRAKE : BYTE := 16#FF;
         END_STRUCT;
END_TYPE;
```

where the element `COLOUR` is an enumeration that can only take the values `red`, `green`, and `yellow`; `SENSOR` is an integer (`INT`); `MOTOR` is a data structure that contains the following elements: `REVOLUTION` with an initial value of 0, `LEVEL` with an initial value of 0; `MAX` with an initial value of `FALSE`; `FAILURE` with an initial value of `FALSE`; and `BRAKE` with an initial hexadecimal value of `FF`.

7.4 Ladder Logic (LL)

The programming language **ladder logic (LL)** comes from the domain of electromechanical relay systems and describes the flow of electricity through a single network representing the POU. This programming language is primarily used for working with Boolean signals.

The ladder network or diagram consists of two vertical tracks and horizontal rungs connecting the vertical tracks. It is assumed that "electricity" flows from the left-hand track to the right-hand track following the rungs. Thus, the ladder network is always read rung by rung from top to bottom and in a given row from left to right, as long as no other order is provided. It is traditional to label all the left-hand rungs with a number. Normally, the numbers are not sequential but increase in units of 5 or 10 in order that additional future rungs can be easily added. Finally, the ladder network is often visually split into two parts: a left-hand part that shows the computations and a right-hand part that shows the storing or using of the variables. This convention makes reading the ladder network easier.

7.4.1 Components of Ladder Logic

Table 7.5 shows the components of ladder logic.

Table 7.5 Components of ladder logic

Name	Symbol	Commentary
Rung	|——|	Read from left to right
Open contact	VarName —| |—	Copies the value from left to right, when the value of the variable `VarName` is `TRUE`; otherwise, `FALSE` is copied
Closed contact	VarName —|/|—	Copies the value from left to right, when the value of the variable `VarName` is `FALSE`; otherwise, `TRUE` is copied
Positive-transition-sensing contact	VarName —|P|—	Copies the value from left to right if and only if a `FALSE` → `TRUE` transition in the variable `VarName` is detected; otherwise, `FALSE` is copied
Negative-transition-sensing contact	VarName —|N|—	Copies the value from left to right if and only if a `TRUE` → `FALSE` transition in the variable `VarName` is detected; otherwise, `FALSE` is copied
Coil[3]	VarName —[ ]—	Copies the value on the left into the variable `VarName`
Negated coill[3]	VarName —[/]—	Copies the negated value of the left into the variable `VarName`
Set coil[3]	VarName —[S]—	Copies `TRUE` to variable `VarName`, if the left link is `TRUE`; otherwise, no change
Reset coil[3]	VarName —[R]—	Copies `FALSE` to variable `VarName`, if the left link is `TRUE`; otherwise, no change
Positive-transition-sensing coil[3]	VarName —[P]—	Saves `TRUE` to the variable `VarName` if and only if a `FALSE` → `TRUE` transition is detected on the left link; otherwise, no change in the variable
Negative-transition-sensing coil[3]	VarName —[N]—	Saves `TRUE` to the variable `VarName` if and only if a `TRUE` → `FALSE` transition is detected on the left link; otherwise, no change in the variable

(continued)

[3] The value on the left is always transferred to the right-hand side.

Table 7.5 (continued)

Name	Symbol	Commentary
Set-Reset Block	SR: S Q1 R	This block combines the functions of the set and reset coils
Return	├⟨RETURN⟩	Exits the POU and returns to the calling POU
Conditional return	├[tnw]─⟨RETURN⟩	If the left link `tnw`[4] is `TRUE`, exit the POU and return to the calling POU; otherwise, no meaning
Jump	├>> NAME	Jump to the network with the given `NAME`
Conditional jump	├[tnw]─>> NAME	If the left link `tnw4` is `TRUE`, jump to the network with the given `NAME`
Label	LABEL ├─	Shows the name for part of a network

Figure 7.3 shows how the typical Boolean operators can be implemented in ladder logic. The `AND` operator is implemented by placing the two contacts in series, while the `OR` operator places the two contacts in parallel. This follows from the observation that the electricity flows from left to right. For an `AND` operation, we need both contacts to be true for the electricity to flow. This implies that both need to be in series. On the other hand, for the `OR` operation, electricity can flow through either of the two paths. Therefore, the contacts should be placed in parallel.

Functions and Ladder Logic

Since ladder logic was originally developed for logical (or Boolean) systems, the implementation and running of complex functions using the simple ladder-logic components can be difficult. For this reason, it is possible to define a function block that is programmed using another programming language. This function block always contains two Boolean variables (`EN` and `ENO`) as well as all the other required parameters. The Boolean input variable `EN` (enable in) determines if the function will be called. If `EN` is `TRUE`, then the function is called. The Boolean output variable `ENO` (enable out) determines if the programme completed successfully. It takes the value `TRUE`, if no errors occurred. Figure 7.4 shows such an implementation using ladder logic.

---

[4] The variable tnw represents a Boolean variable that determines if the given link should be performed.

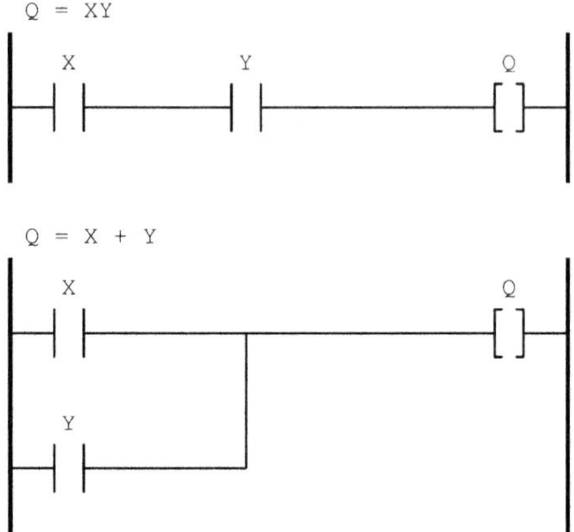

Fig. 7.3 (Top) AND and (Bottom) OR in ladder logic

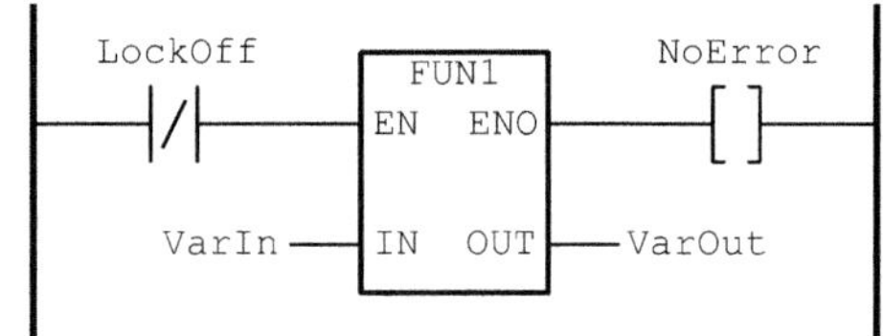

Fig. 7.4 Calling a function in ladder logic

7.4.2 *Examples of Using Ladder Logic*

Example 7.1: Ladder Logic for a Boolean Function
Please write the corresponding ladder diagram for the following Boolean function:

$$Q = XY + XZ \quad + YZ$$

Solution

When we want to create the ladder-logic diagram, we should first convert the Boolean function into a minimal form. In our case, we can easily convert it to

$$Q = XY + (X + Y)Z.$$

In the ladder-logic diagram, we will require (at least) one row for each term that is separated by a "+", since each row corresponds to an AND Term. Thus, in our example, we will require 3 rows. All the values are not negated. Thus, the open contact will be used for all values. In order to save the resulting function value in Q, we require a coil. Figure 7.5 shows the ladder-logic diagram for this example.

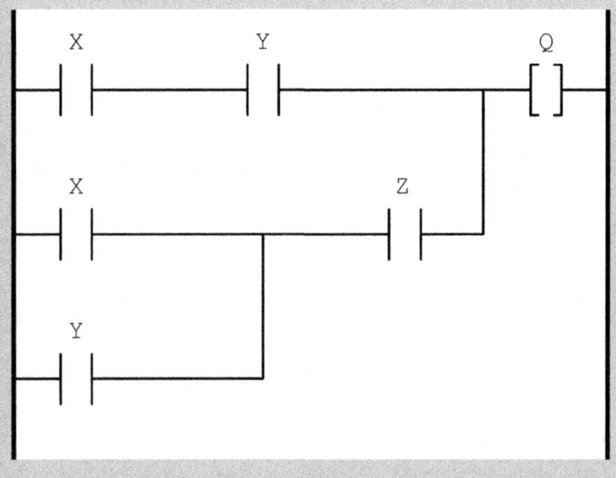

Fig. 7.5 Ladder logic for Example 7.1

Example 7.2: Ladder Logic for a Recipe

Consider a tank that needs to be filled with two components and then mixed before being sent on its merry way. The recipe is:

1. Once the start button is pressed and both level sensors (`L1` and `L2`) read `FALSE`, turn on Valves 1 and 2 (`V1` and `V2`). Go to Step 2.
2. Once the fluid reaches level sensor 2 (`L2`), that is, it reads `TRUE`, close both valves and turn on the mixer (`M1`) for five minutes. Go to Step 3.
3. Turn off the mixer and wait one minute. Go to Step 4.
4. Open the bottom valve (`V3`) and let the fluid drain. Go to Step 5.
5. Once the bottom level sensor 1 (`L1`) reads `FALSE`, that is, the tank is empty, close the bottom value. Go to Step 1.

Implement this recipe using ladder logic.

Solution

In order to implement a recipe in ladder logic, we will need to define Boolean variables that monitor in which step we are. In this example, since we have five steps, it makes sense to define the Boolean step variables as `S1`, `S2`, `S3`, `S4`, and `S5`. `S1`will be initialised as `TRUE`, while all other variables will be initialised as `FALSE`. At the end of each step, the current step variable will be reset (to zero) and the next step variable will be set (to one).

Turning on and off valves will be implemented using the set and reset coils, while the delay and timer will be implemented using the built-in time function (TON). Rung numbers have been added in multiples of 5. Finally, it can be noted that in this particular recipe each set has a corresponding reset. This is a feature of many recipes that will be made explicit in the PLC programming language sequential function charts. Figure 7.6 shows the final programme using ladder logic.

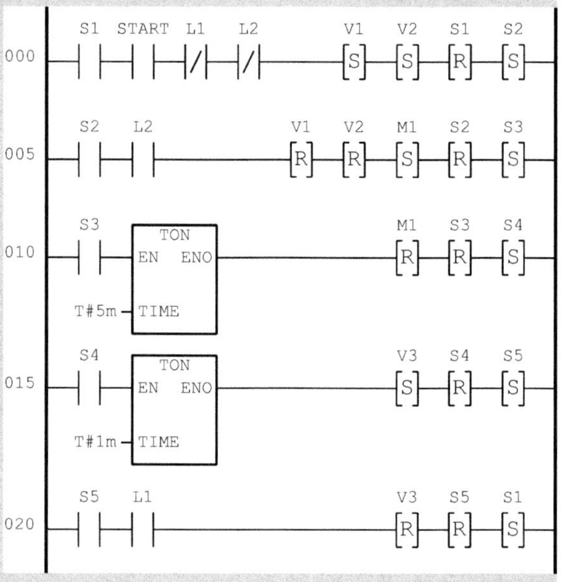

Fig. 7.6 Ladder logic for Example 7.2

7.4.3 Comments

Ladder logic is perfect for large Boolean networks. However, it is not appropriate for sequence or recipe programming. In ladder logic, it is not recommended to use jumps, since these can lead to inconsistencies during runtime.

7.5 Instruction List (IL)

Instruction list (IL) is a language similar to assembly language. This language is a simple list of instructions, each of which is placed on a separate line. Each instruction has the following form:

`Label:` `Operator/Function Operand` (list) *`Comments`*

where `Label` names a given row of instructions and along with the colon can be ignored; the `Operator` or `Function` is an operator in instruction list or a function; `Operand` is zero, one, or more constants or variables for the operator or input parameters for the functions that are separated by commas; and `Comments` is an optional field that provides additional information about what the given instruction row does. All fields in *italics* are optional.

7.5.1 Universal Accumulator

Most assembly languages have a physical accumulator in the processor, that is, a value is loaded into this accumulator, additional values are added, subtracted, and then stored. Instruction list also has such an accumulator that is often called the **current result** (CR). However, it is not a memory area with permanently defined register lengths. Instead, the statement-list compiler ensures that an abstract accumulator of any memory width is available that depends on the data type to be processed. In contrast to other assembly languages, there are no special processor status bits. The result of comparisons is written as a Boolean `0` or `1` in this accumulator. Conditional jumps and calls use the current value stored in the accumulator to evaluate the jump requirements. The current result can be of any general data type, derived data type, or function-block type. The data width of the current result (number of bits) does not matter. Instruction list requires that two consecutive operations be type compatible, which means that the data type of the current result must be compatible with the following statement.

Table 7.6 shows the modifications of the current result by different operator groups. *Unchanged* means that an instruction passes the result of the previous instruction on to the following one without having made any changes to the value and type. *Undefined* means that the subsequent operation cannot use the current result. The first command of a function block (which is called using `CAL`) may only be a load `LD`, jump `JMP`, function block call `CAL`, or return `RET`, since these commands do not require a current result.

The IEC 61131-3 standard itself does not define any operator groups. In fact, the behaviour and evaluation of the current result are only partially described in the standard. With operations such as `AND`, the type and content of the current result are intuitively clear both before and after execution. However, the standard leaves open the question of how the current result is defined after an unconditional jump.

Table 7.6 Changes in the current result for different operator groups

Effect of the operator group on the current result	Abbreviation	Example operators
Create	C	LD
Process	P	GT
Unchanged	U	ST; JMPC
Undefined	–	Unconditional CAL block call, since the following instruction must reload the current result, because it does not have a clearly defined value after the function block has been returned

7.5.2 *Operators*

Table 7.7 shows the operators defined in instruction list. The modifiers are defined as follows:

Table 7.7 Operators in instruction list

Operator	Acceptable modifiers	Operand	Definition
LD	N	ANY	Load
ST	N	ANY	Save
S		BOOL	Set
R		BOOL	Reset
AND/&	N, (	ANY	Boolean AND
OR	N, (	ANY	Boolean OR
XOR	N, (	ANY	Boolean Exclusive OR
ADD	(	ANY	Addition
SUB	(	ANY	Subtraction
MUL	(	ANY	Multiplication
DIV	(	ANY	Division
GT	(	ANY	>(Greater than)
GE	(	ANY	≥(Greater than and equal to)
EQ	(	ANY	=(Equal to)
NE	(	ANY	≠(Not equal to)
LT	(	ANY	<(Less than)
LE	(	ANY	≤(Less than and Dollar sign to)
JMP	C, (	LABEL	Jump to LABEL
CAL	C, (	NAME	Call function NAME
RET	C, (		Return from a function call
)			Take the last deferred instruction

- `N`: Negation;
- `(`: Nesting levels using brackets;
- `C`: Conditional performance of an operator when the current result equals `TRUE`.

The modifiers are written together with the operator, for example, `ANDN` is a negated `AND`.

The current result can be linked with the result of an instruction sequence using the bracket operators. When the modifier `(` appears, the associated operator, the value of the current result and the data type of the current result are cached. The data type and value of the following line are loaded into the current result. When the `)` operator appears, the cached values and data type based on the operator and modifiers are linked with the current result. The result is then stored as the current result. Expressions in brackets can be nested.

7.5.3 Functions in Instruction List

In instruction list, functions are called by giving their name. The parameters can be passed using two different approaches: **actual parameters** or **formal parameters**. With actual parameters, the first parameter of a function is the current result that was previously loaded. Therefore, the first operand after the function name is the second function parameter in this variant. With formal parameters, all parameters are explicitly defined. Table 7.8 shows the two possibilities.

A function has at least one output parameter (function value) that is returned using the current result. Should the function have additional output parameters, these can be returned using parameter assignments. A call without formal parameters takes place in one line. The original order of the output declarations must be observed. In the case of formal parameter assignments, this is done line by line, followed by a closing bracket on its own line. With formal parameter assignment, the output parameters are marked with `=>`, e.g. `ENO => ErrorOut` means that `ENO` is an output value and is saved in the `ErrorOut` variable. The programming system assigns the function value to a variable with the function name. This name is declared automatically and does not have to be specified separately by the user in the declaration section of the calling block.

Table 7.8 Two possibilities for calling the function `LIMIT(MN, IN, MX)`

Possibility 1: Actual parameters	Possibility 2: Formal parameters

`LD 1`
`LIMIT 2, 3`
	`LIMIT(`
	`MN := 1,`
	`IN := 2,`
	`MX := 3`
	`)`

Table 7.9 Three methods for calling the function block TIME1(IN, PT) with output variables Q and ET

Method 1	Method 2	Method 3

CAL TIME1(
  IN := Free,
  PT := t#500_ms,
  Q => Out,
  ET => VALUE
)
	LD t#500_ms
	ST TIME1.PT
	LD Free
	ST TIME1.IN
	CAL TIME1
	LD TIME1.Q
	ST Out
	LD TIME1.ET
	ST VALUE
		LD t#500_ms
		ST TIME1.PT
		LD Free
		IN TIME1
		LD TIME1.Q
		ST Out
		LD TIME1.ET
		ST VALUE

7.5.4 *Calling Function Blocks in Instruction List*

Function blocks are called using the CAL operator. The IEC 61131-3 standard provides three methods for calling function blocks:

(1) calling a function block with a bracketed list of input and output parameters;
(2) calling a function block with previously loaded and saved input parameters; and
(3) calling a function block implicitly by using the inputs as operators.

Method 3 can only be used for standard function blocks, since the programming system must use the name of the function block inputs as operators. This method cannot be used for the output parameters. Table 7.9 shows the three methods for calling the function block TIME1(IN, PT) with output variables Q and ET.

7.5.5 *Examples*

Example 7.3: Example of the Computation of the Current Result
For the given instructions, follow how instruction list uses and updates the different registers. Values in grey are the default initial values.

Instruction	(* Comments *)	CR	X1	X2	W1	FB1.i1	FB1.o1
	Initial values:	"STR"	0	1	10	1	103
LDN X1	Load the negated value in X1	**1**	**0**	1	10	1	103
AND X2	CR AND X2 → CR	**1**	0	**1**	10	1	103
S X1	If CR=1 then: 1→X1	**1**	**1**	1	10	1	103

Instruction	(* Comments *)	CR	X1	X2	W1	FB1.i1	FB1.o1
R X2	If CR=1 then: 0→X2	**1**	1	**0**	10	1	103
JMPCN Lab1	Jump to Lab1 if CR=0	**1**	1	0	10	1	103
LD W1	Load the value in W1→CR	**10**	1	0	**10**	1	103
MUL W1	CR * W1→CR	**100**	1	0	**10**	1	103
SQRT	**Function**: sqrt(CR) →CR	**10**	1	0	10	1	103
ST FB1.i1	Save CR→FB1.i1	**10**	1	0	10	**10**	103
CAL FB1	**FB1**: internal computation	10	1	0	10	10	**112**
LD FB1.o1	Load the output o1 from FB1→CR	**112**	1	0	10	10	112
ST W1	Save CR→W1	**112**	1	0	**112**	10	112
GT 90	(CR>90)?→CR	**1**	1	0	112	10	112

Example 7.4: Writing an Instruction-List Programme

Please write the instruction-list programme for the following function:

$$Q = XY + XZ \quad + YZ$$

Solution

The simplest approach to writing the instruction-list programme is to go from left to right and write the corresponding instruction. The programme is then

```
LD X
AND Y
OR( X
AND Z
)
OR( Y
AND Z
)
ST Q
```

Comments

Instruction list is perfect for large Boolean networks. However, like with ladder logic, it is inappropriate for sequence or recipe programming. Instruction list has no complex programming flow concepts, such as loops. Finally, in instruction list, jumps should be avoided as they can lead to inconsistencies during runtime.

7.6 Function-Block Language (FB)

The function-block language is a graphical model with Boolean operators and complex functionality represented using blocks. Motivated by electronic circuits, the signals flow between the blocks and are converted from Boolean inputs into Boolean outputs. Although this language is based on electrical circuits, the current standard has adopted different representations for the blocks symbolising `AND` and `OR`.

7.6.1 Blocks Used in the Function-Block Language

Table 7.10 shows the blocks used in the function-block language.

Inputs and outputs are defined upon first use. Figure 7.7 shows an example of a function-block diagram with common components. The exact method for labelling the variables is not specified in the standard; any reasonable method can be used.

7.6.2 Feedback Variable

The function-block language allows an output to be used as the new input of a network. Such a variable is called a **feedback variable**. The first time such a situation is encountered the initial value is used; afterwards, the last value is used. Figure 7.8 shows an example with feedback.

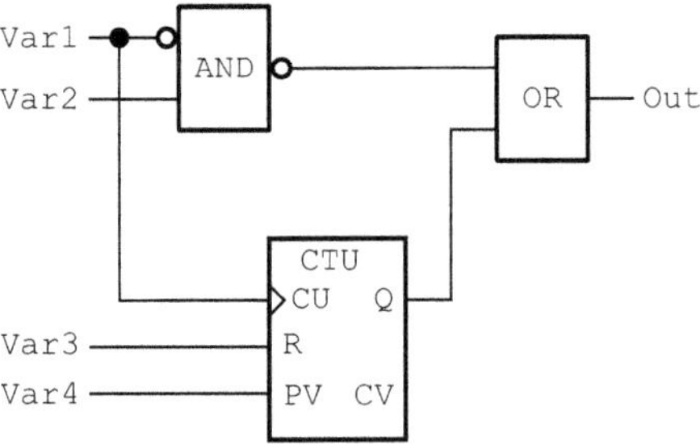

Fig. 7.7 Diagram using the function-block language

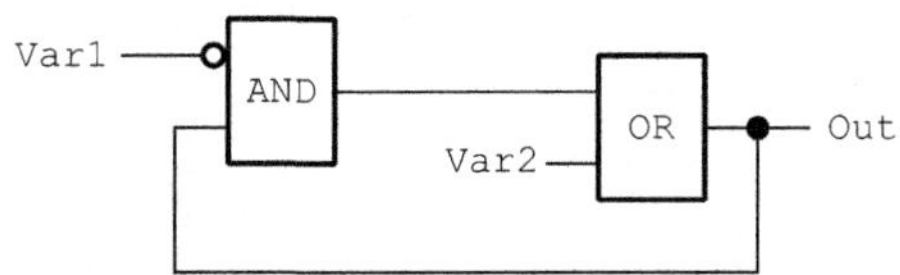

Fig. 7.8 Feedback in the function-block language

Table 7.10 Blocks used in the function-block language

Name	Symbol	Comments
AND	AND	
OR	OR	
Negation	○	Negation is placed where the signal enters or leaves a block
Link	●	When two connections coincide, the value is to be transferred
General block	NAME	There are as many input and output ports as required by the block
Rising edge	>	This is placed within the function block next to the corresponding variable in order to show a rising-edge transition
Falling edge	<	This is placed within the function block next to the corresponding variable in order to show a falling-edge transition
Return	—<RETURN>	Leaves a POU and returns to the calling POU
Conditional return	—[snw]—<RETURN>	If the left connection `snw`[5] `is TRUE`, then exit the POU and return to the calling POU; otherwise ignored
Jump	—>> NAME	Jump to the network with the given identifier `NAME`
Conditional jump	—[snw]—>> NAME	If the left connection `snw`[6]`is TRUE`, jump to the network with the given identifier `NAME`; otherwise ignored

[5] The variable snw represents a subnetwork that returns a Boolean value.

[6] The variable snw represents a subnetwork that returns a Boolean value.

7.6.3 Example

Example 7.5: Creating the Diagram for the Function-Block Language
Please create the diagram using the function-block language for the following Boolean function:

$$Q = XY + XZ + YZ$$

Solution

Figure 7.9 shows the solution.

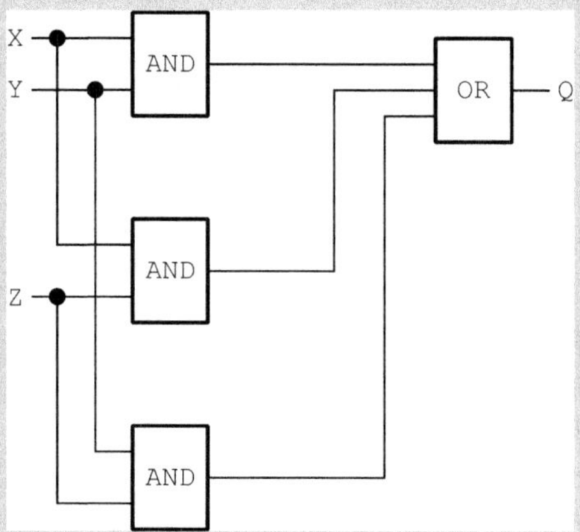

Fig. 7.9 Function Q = XY + XZ + YZ in the function-block language

7.6.4 Comments

Function-block language should be used in cases where there is signal flow between different components, for example, in closed-loop control. This language contains no complex programme flow concepts, such as loops.

7.7 Structured Text (ST)

Structured text (ST) is a further text-based programming language in the IEC 61131-3 standard. It is referred to as a high-level programming language, since assembly-language commands are not used in structured text, but powerful constructs are built using more abstract commands. In structured text, the solution to a particular programming task is broken down into individual steps (instructions). An instruction is used to calculate and assign values, control the flow of commands, or to call or exit a POU.

7.7.1 Commands in Structured Text

Table 7.11 shows the commands in structured text. Each command must be separated by a semicolon, that is, unlike previous text-based languages, multiple commands can be placed on a single line.

The `IF` command has many different forms. The obligatory components are the `IF`, `THEN`, and `END_IF` commands. The `IF` command opens the `IF` block, while the `END_IF` closes it. The other possibilities are the `ELSE` and `ELSIF` commands. The `ELSE` block is only performed if all previous expressions have been evaluated as `FALSE`. The `ELSIF` block is only performed if all previous expressions have been evaluated as `FALSE` and the corresponding Boolean expression is `TRUE`. This command can be repeated as often as desired. This gives the following possibilities:

(1) `IF` expression `THEN` code block; `END_IF`;
(2) `IF` expression `THEN` code block; `ELSE` code block; `END_IF`;
(3) `IF` expression `THEN` code block; `ELSIF` expression `THEN` code block; `ELSIF` expression `THEN` code block; `ELSE` code block; `END_IF`.

The `FOR` command also has many different possibilities. The obligatory parts are `FOR`, `TO`, `DO`, and `END_FOR`. The `FOR` command opens this code block, and the `END_FOR` command closes it. The `TO` command specifies the end value, and the `DO` command closes the declaration line of the `FOR` command. The optional part is the `BY` command that gives the step value. If no step value is given using the `BY` command, then it is assumed that the step is one. The two possibilities for the `FOR` command are:

(1) `FOR Counter :=`expression `TO` expression `BY` expression `DO` code block; `END_FOR;`
(2) `FOR Counter :=`expression `TO` expression `DO` code block; `END_FOR;`

Table 7.11 Commands in structured text

Keyword	Description	Example	Comments
:=	Assignment	d := 10;	Assigns the value on the right to the variable on the left
:=,=>	Calling and Using Function Blocks	FBNAME(Part1 := 10, Part3 => OUT);	Calls another POU of type function block and assigns the required parameters: :=for inputs and => for outputs
RETURN	Return	RETURN;	Returns to the calling POU

IF ... THEN
ELSE,
ELSIF
END_IF
	Branching	IF A>100 THEN
		f := 1;
		ELSIF d=e THEN
		f := 2;
		ELSE
		f := 4;
		END_IF
			Selection of alternatives using Boolean logic

CASE ... OF ELSE END_CASE	Multiple Selection	CASE f OF 1: g := 11; 2: g := 14; ELSE g := -1; END_CASE;	Selects a code block based on the value of the expression f
FOR ... TO ... BY ... DO END_FOR	FOR Loop	FOR h := 1 TO 10 BY 2 DO f[h/2] := h; END_FOR;	Repeated running of a code block with start and end conditions

WHILE ... DO
END_DO
	WHILE Loop	WHILE m > 1 DO
		n := n/2;
		END_WHILE;
			Repeated running of a code block with end conditions

REPEAT ... UNTIL
END_REPEAT
	REPEAT Loop	REPEAT
		i := i*j;
		UNTIL i > 10,000
		END_REPEAT;
			Repeated running of a code block with end conditions

EXIT	Breaking a loop	EXIT;	Immediate exit from a loop
;	Command delimiter		Shows the end of a command

7.7.2 *Operators in Structured Text*

Table 7.12 shows the priority of the operators in structured text.

Table 7.12 Operators and their priority in structured text

Operator	**Description**	**Priority**
`(...)`	Brackets	**Highest**
`Function(...)`	Function Evaluation	
`**`	Exponentiation	
`-, NOT`	Negation, Boolean Complement	
`*, /`	Multiplication, Division	
`MOD`	Modulo	
`+, -`	Addition, Subtraction	
`>, <, ≤, ≥`	Comparison Operators	
`=`	Equality	
`<>`	Inequality (≠)	
`AND, &`	Boolean AND	
`XOR`	Boolean Exclusive OR	
`OR`	Boolean OR	**Lowest**

7.7.3 Calling Function Blocks in Structured Text

In structured text, function blocks are called by their name with the required parameters placed in brackets. The := is used to assign the value of the actual parameters, while => is used to assign any output variables. Since the assignments must be explicit, the order of the assignments is irrelevant. If a parameter is not initialised during the call, then the initial value or the last value will be used.

7.7.4 Example

Example 7.6: Structured Text
Write the structured-text programme for the following system. As shown in Fig. 7.10, a tank can be filled using valves. The weight of the tank is determined using a scale. The function block monitors the tank weight to determine if the tank is full, empty, or in between the two states. The block gives a single command that can have four values:

1. fill the tank;
2. stop filling the tank;
3. start the stirrer;
4. empty the tank.

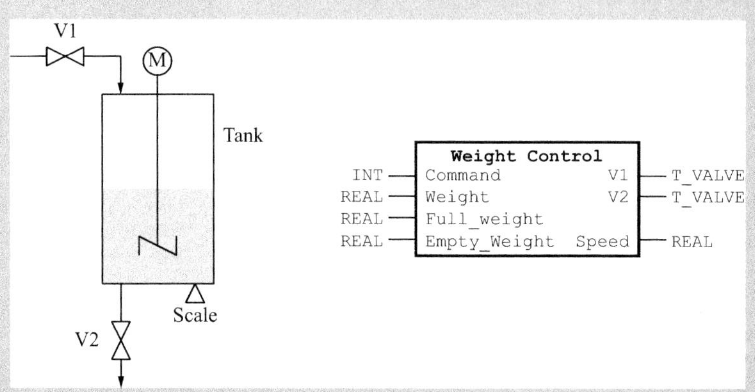

Fig. 7.10 Tank system for the structured-text example

As necessary, the appropriate values are opened or closed to control the tank level. The stirrer can only work if the tank is full; otherwise, the command is ignored.

Solution

```
(*Tank States*)
TYPE T_STATE: (FULL,NOT_EMPTY,EMPTY); END_TYPE;
(*Valve States*)
TYPE T_VALVE: (ON,OFF); END_TYPE;

FUNCTION_BLOCK Weightcontrol
VAR_IN
      Command: INT;
       Weight: Real;
       Full_weight,Empty_weight:  Real;  (*Same  data  type
declared on one line*)
END_VAR;
VAR_OUTPUT
       V1: T_VALVE := OFF;
       V2: T_VALVE := OFF;
       Speed: REAL := 0.0;
END_VAR;
VAR (*Internal Variables*)

       State: T_STATE :=EMPTY;

END_VAR;
(*Determine the Tank State: Compare the full and empty weights *)
```

```
IF Weight >= Full_weight THEN
      State := FULL;
ELSIF Weight <= Empty_weight THEN
      State := EMPTY;
ELSE
      State := NOT_EMPTY;
END_IF;
(*Implementation of Commands: 1-Fill, 2-Stop, 3-Stir, 4-Empty*)
CASE Command OF
      1: V2 := OFF;
      V1:=SELECT(G := State= FULL,IN0 := ON,IN1 := OFF);
(*ON only if G is false*).
      2: V2 := OFF;
      V1 := OFF;
      4: V1 := OFF;
      V2 := ON;
END_CASE;
(*Stirrer Speed*)
Speed:= SELECT(G := Command=3; IN0 := 0.0; IN1 : = 100.0);
END_FUNCTION_BLOCK;
```

7.7.5 Comments

Structured text has the following advantages (especially in comparison with instruction list): compact formulation of the programming tasks, clear programme structure, and powerful constructs to control the flow of commands. However, it has the following disadvantages. The conversion of the programme into machine code cannot be directly influenced since it is performed using a compiler. As well, the higher abstraction level brings with it a loss of efficiency, that is, the translated programme is longer and slower.

7.8 Sequential-Function-Chart Language (SFC)

The **sequential-function-chart language(SFC)** is the third graphical method for programming a PLC. It consists of a sequence of steps and transitions that implement predefined tasks and provide a visual overview of the process.

Fig. 7.11 Steps in sequential function charts: (left) general step and (right) initial step

7.8.1 Steps and Transitions

A **step** is either active or inactive. It consists of a number of instructions that are performed as long as the given step is active. A **transition** uses a Boolean expression to determine when a step becomes inactive. Connections with a predefined direction describe which step or steps should be activated next.

Steps are shown using a rectangle. Figure 7.11 shows two possible forms for denoting a step. The general step block, which is shown in Fig. 7.11 (left), is a simple rectangle with a single border. The initial step block, which is shown in Fig. 7.11 (right), is a simple rectangle with a double border.

The transition conditions are marked with a horizontal line with identifiers. The Boolean description, often called a **guard**, shows the requirements for a transition to occur. The guard can be written in any of the other PLC languages. Figure 7.12 shows some possibilities. The most frequently used PLC languages for describing transitions are structured text, function-block language, or ladder logic. Often, the guard is ignored.

When the POU is called, the specially marked initialisation step is made active. All the assigned instructions will be performed. When the transition conditions become TRUE, the initialisation step will be made inactive and the next step will be activated. When a transition occurs, the active attribute (often called a token) is passed from the current active step to its successor. This active attribute wanders through the individual steps, multiplying for parallel branches and coming together again. It may not be completely lost nor may it be in uncontrolled distribution, which occurs when multiple tokens are present in a single step.

7.8.2 Action Blocks

An **action block** shows the details of what must be accomplished in a given step. Each action block is always assigned to a particular step. It is not necessary to include the action block. No more than two action blocks can be associated with a given step.

Each action block has four components, which are shown in Fig. 7.13:

(a) **Qualifier** that consists of an acceptable tag (see Table 7.13).
(b) **Action Name** that briefly describes the action.
(c) **Indicator Variable** that provides which PLC variable is to be used.
(d) **Process Description** that describes, in an appropriate (human) language, the action (optional).

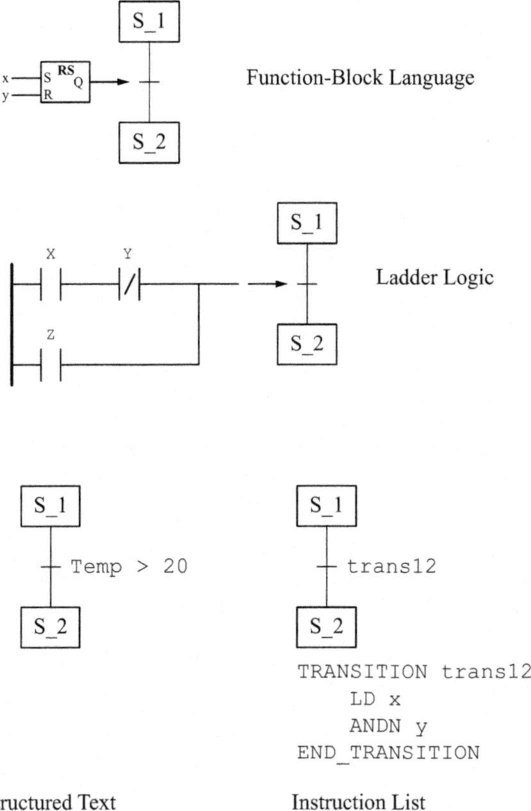

Fig. 7.12 Transitions conditions in different PLC languages

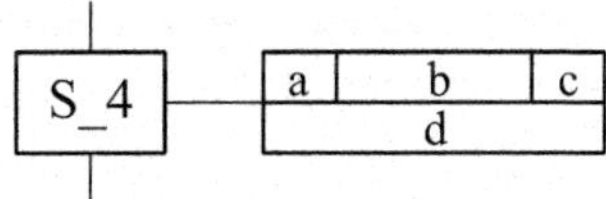

Fig. 7.13 Components of an action block: a: qualifier; b: action name; c: indicator variable; d: process description

Each component has a specified location in the action block. The qualifier must be selected from a small group of abbreviations that describe how the step is to be performed. Table 7.13 shows the possible qualifiers.

Table 7.13 Qualifiers in sequential function charts

Qualifier	Short name	Description
N	Not saved	Means that as soon as the step has ended, the action is stopped
S	Saved	Means that the action continues with the given values until it is stopped
R	Reset	Means that a saved action is stopped
L	(Time) Length	Means that the action lasts for the specified time *T*
D	Delay	Means that the action is delayed by the specified time before being implemented
P	Pulse	Means that the action occurs for a very short period of time
DS	Delayed and saved	Means that the action is first delayed and then saved
SD	Saved and delayed	Means that the action is first saved and, then after the delay, if the step is still active, implemented
SL	Saved and time limited	Means that a time-limited action is saved

7.8.3 Sequential Function Charts

As the name suggests, in the sequential-function-chart language, a chart is designed to show how the individual components are linked together. This chart is built using the following rules:

(1) A vertical link connects two steps.
(2) An arrow is used to clarify the direction in which steps occur.
(3) An action block is connected with the step block using a straight horizontal line.

There are two special types of connections: alternative and parallel paths. An **alternative path** is shown using a single horizontal line across all possible alternative ways. This horizontal line is placed at the beginning and end of the region corresponding to the alternative paths. Only one of the paths must end at which point all the other paths will be stopped. Figure 7.14 shows an alternative path. Alternative paths are normally selected from left to right. This is shown in Fig. 7.15. This means that, first, the transition condition for S_2a will be tested. If it is TRUE, this path will be taken and all other paths ignored. If it is FALSE, then S_2b and S_2c (in this order) will be tested. In this case, the paths will be selected in order from S_2a, S_2b, and then S_2c. When a user-defined order is provided, for example, as in Fig. 7.16, then this order is followed. In this example, it means that first S_2c will be tested, followed by S_2b and finally S_2a.

Parallel paths are shown by a double horizontal line across all possible parallel paths. This horizontal line is placed at the beginning and end of the region corresponding to the parallel paths. All parallel paths must end before the process can continue. Figure 7.17 shows an example of a parallel path.

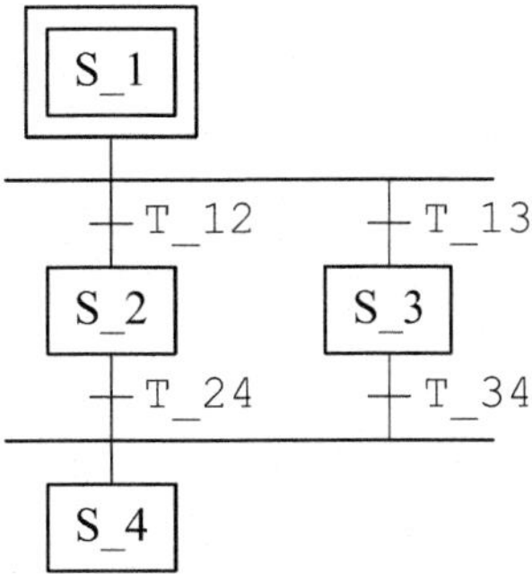

Fig. 7.14 Alternative paths in sequential function charts

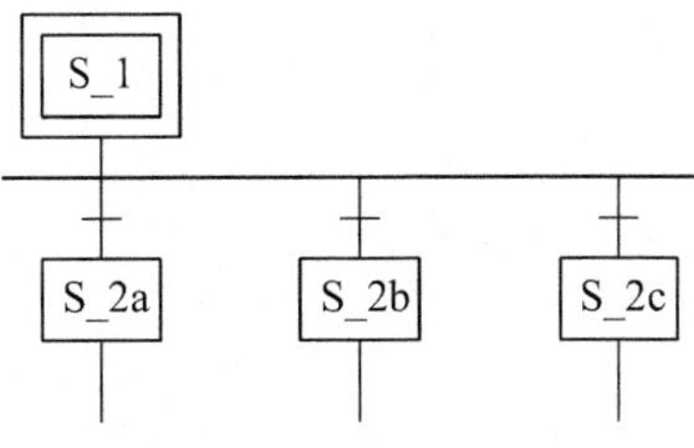

Fig. 7.15 Usual decision order for alternative paths in sequential function charts

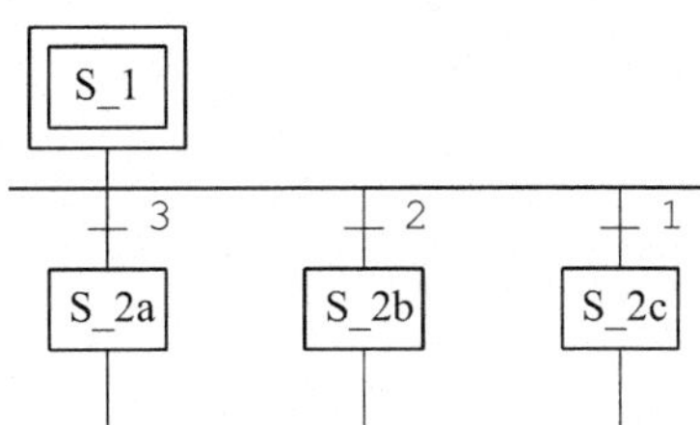

Fig. 7.16 User-defined decision order for alternative paths in sequential function charts

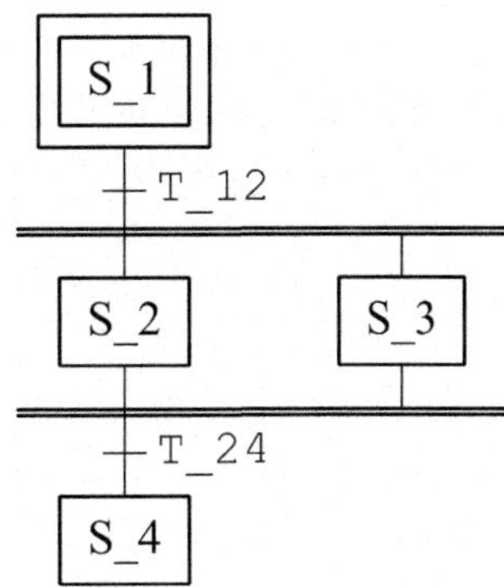

Fig. 7.17 Parallel paths in sequential function charts

7.8.4 Example

Example 7.7 Creating a Sequential Function Chart
Consider a chemical reactor as shown in Fig. 7.18. It is desired to design a PLC to control the process. The objective is to draw the sequential function chart for the process. The desired process can be described as follows.

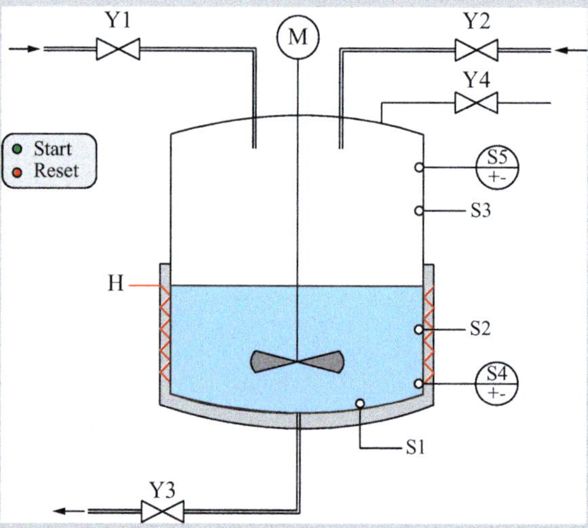

Fig. 7.18 Schematic of the reactor

When the START button has been pressed, the reactor status sensor S1 confirms that the reactor is empty and that the temperature sensor S4 and the pressure sensor S5 give no error messages, then valve Y1 should be opened until level sensor S2 reaches the desired value, that is, it returns a 1. Then, the motor for agitator M should be turned on and valve Y2 opened. When the level sensor S3 reaches the desired value, valve Y2 should be closed. After waiting 5 s, heater H is turned on until the temperature sensor S4 reaches the desired value. During the heating procedure, should the pressure sensor S5 return a high-pressure alarm in the reactor, then the pressure release valve Y4 should be opened until the alarm is lowered. After the heating procedure has been completed, the agitator runs for 10 more seconds, at which point, valve Y3 is opened. Once the reactor is empty, that is, the reactor status given by S1 returns 1, valve Y3 is closed and the process can be restarted.

Solution

Before the solution to the problem is provided, it is useful to consider the general procedure for solving such problems. This procedure can be written as.

1. **Define** all the variables and their values, especially for the Boolean variables.
2. **Write** the process description as a series of steps. Note that a single sentence can be associated with multiple steps, or multiple sentences can be combined into a single step.
3. **Draw** the sequential function chart.

Defining the Variables

The variables are defined as follows.

Input Variable	Symbol	Data Type	Logical Values		Address
Start button	START	BOOL	Pressed	START = 1	E 0.0
Reactor status	S1	BOOL	Tank empty	S1 = 1	E 0.1
Level sensor 1	S2	BOOL	Level 1 reached	S2 = 1	E 0.2
Level sensor 2	S3	BOOL	Level 2 reached	S3 = 1	E 0.3
Temperature sensor	S4	BOOL	Temperature reached	S4 = 1	E 0.4
Pressure sensor	S5	BOOL	High-Pressure Alarm	S5 = 1	E 0.5
Valve 1	Y1	BOOL	Valve open	Y1 = 1	A 4.1
Valve 2	Y2	BOOL	Valve open	Y2 = 1	A 4.2
Release valve	Y3	BOOL	Valve open	Y3 = 1	A 4.3
Pressure release valve	Y4	BOOL	Valve open	Y4 = 1	A 4.4
Heater	H	BOOL	Heater on	H = 1	A 4.5
Agitator motor	M	BOOL	Motor on	M = 1	A 4.6

Figure 7.19 shows the sequential function chart along with the corresponding text, so that it is clear how each of the components was created.

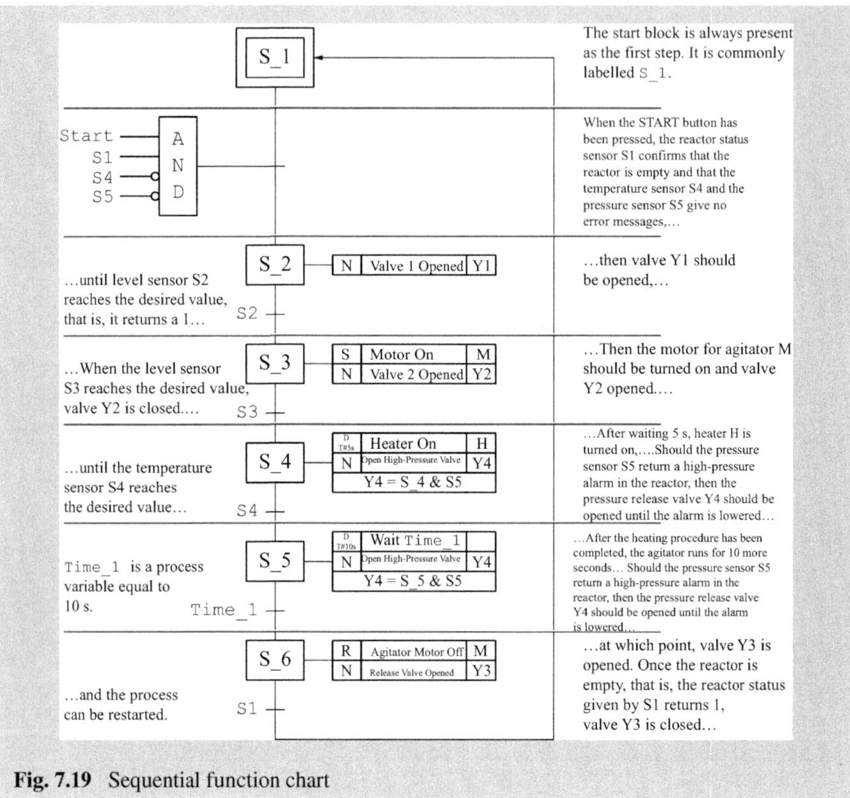

Fig. 7.19 Sequential function chart

7.8.5 *Validity of a Sequential Function Chart*

With sequential function charts, it is easy to create complex networks. However, the question remains if these networks are truly valid and result in proper operation.

Although there exists a method to determine the validity of a sequential function chart, it is not guaranteed that it will find all valid networks. The proposed method only determines if the network is valid. If the method fails, there is no guarantee that the chart is truly invalid. The procedure is:

(1) Replace all step-transition-step by a single step.

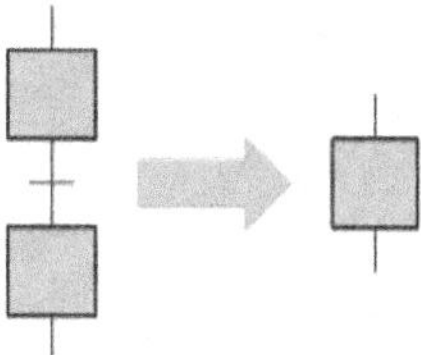

(2) Replace all transition-step-transition by a single transition.

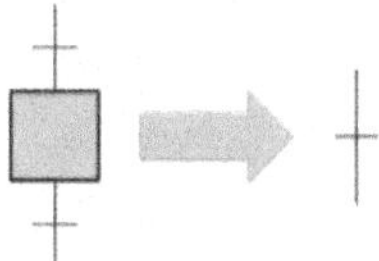

(3) Replace two parallel transitions by a single transition.

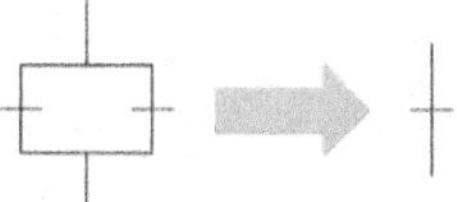

(4) Replace two parallel steps by a single step.

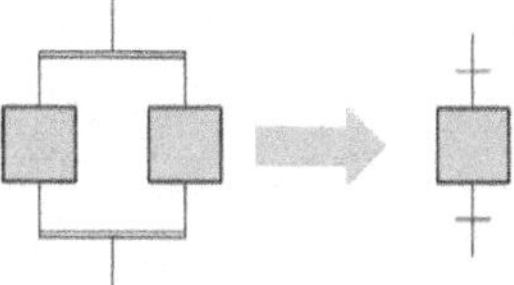

(5) Remove any self-transitions.

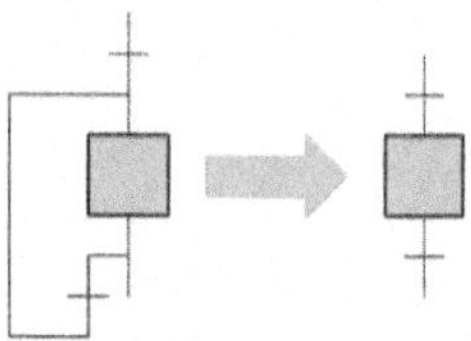

The validity is then determined using the following rule: a sequential function chart is valid when the chart can be reduced to a single, implementable step.

Example 7.8 Determining the Validity of a Network
Determine the validity of the sequential function chart shown in Fig. 7.20.

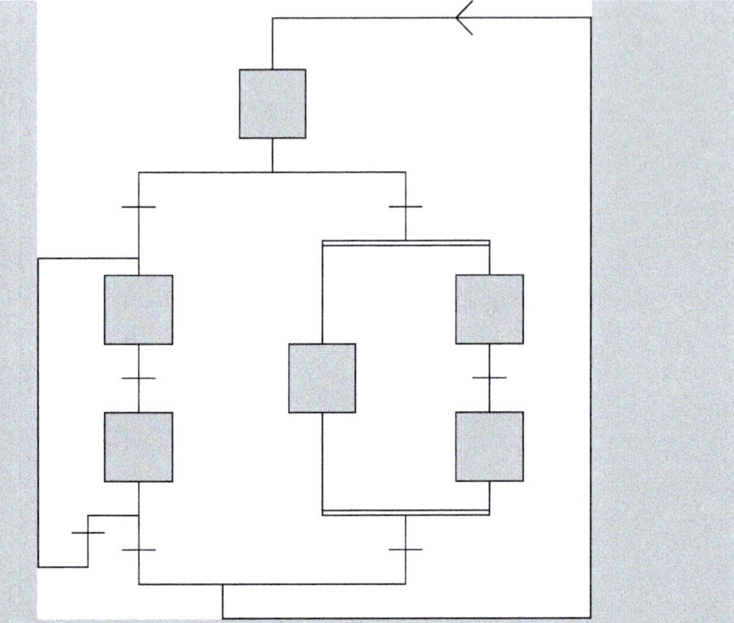

Fig. 7.20 Sequential function chart that is to be checked for its validity

Solution

The method is followed until no further reductions can be produced. The first step is to apply Rule #1 to give Fig. 7.21.

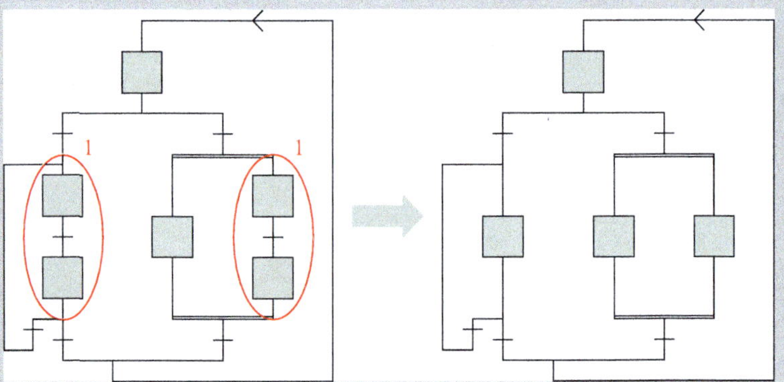

Fig. 7.21 First reduction

Then, Rules #4 and 5 can be applied to give Fig. 7.22.

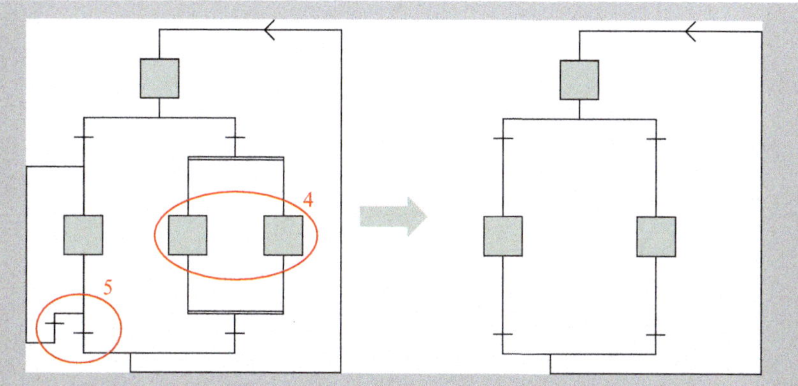

Fig. 7.22 Second reduction

Then, Rule #2 is applied twice to give Fig. 7.23.

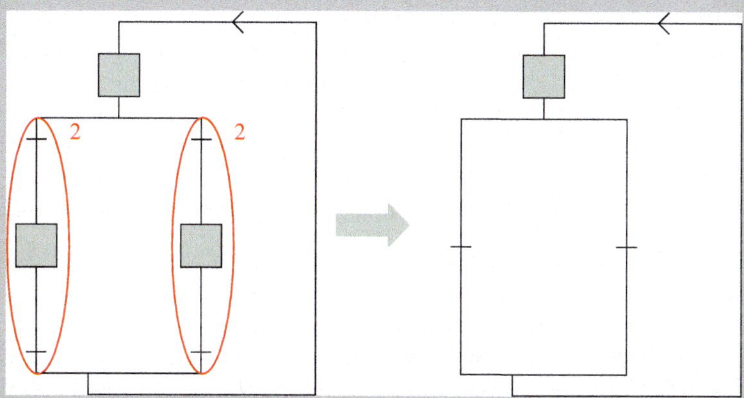

Fig. 7.23 Third reduction

The last step involves applying Rule #3 to give Fig. 7.24.

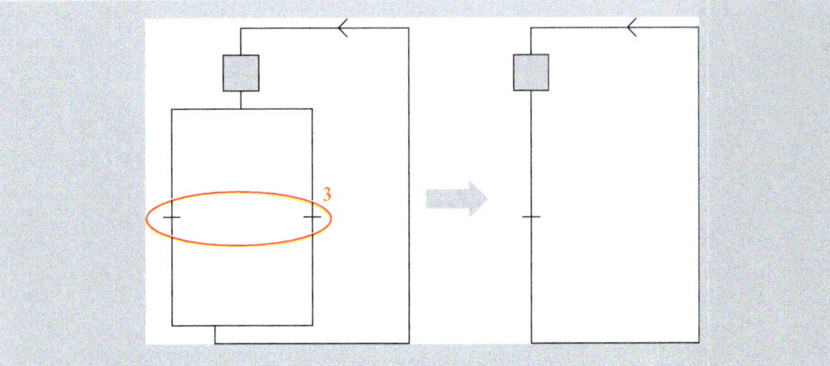

Fig. 7.24 Fourth and final reduction

The reduced chart is obviously valid, since it contains only a single step and a single transition. Thus, we can conclude that the proposed sequential function chart is in fact valid.

7.9 Further Reading

The following are additional resources on the topics mentioned in this chapter:

(1) K.-H. John and K. Tiegelkamp (2009). *IEC 61131-3: Programming Industrial Automation Systems: Concepts andProgramming Languages, Requirements for Programming Systems, Decision-Making Aids* (Second Edition), Berlin, Germany: Springer-Verlag.

7.10 Chapter Problems

Problems at the end of the chapter consist of three different types: (a) Basic Concepts (True/False), which seek to test the reader's comprehension of the key concepts in the chapter; (b) Short Exercises, which seek to test the reader's ability to compute the required parameters for a simple data set using simple or no technological aids; and (c) Computational Exercises, which require not only a solid comprehension of the basic material, but also the use of appropriate software.

7.10.1 Basic Concepts

Determine if the following statements are true or false and state why this is the case.

(1) Global variables can only be used in a single resource.
(2) A non-pre-emptive task must always be completed before another task can be started.
(3) `1Prog12` is a valid function name.
(4) According to the IEC standard, `TUI_124` and `TUI_12456` are equal.
(5) `T#4d4.2h` represents 4 days and 4.2 h.
(6) `LE` means *less than*.
(7) A task with a priority value of 3 has the highest priority.
(8) In the function-block language, a double arrow denotes a jump between two parts of the function-block network.
(9) In the function-block language, we can use `&`.
(10) In ladder logic, a coil can be used to save a value.
(11) In ladder logic, we can create feedback loops.
(12) In instruction list, the statement `SN HIPPO` is a valid one.
(13) In instruction list, the brackets `()` delay the implementation of an instruction.
(14) In structured text, `OR` has a higher priority compared to `AND`.
(15) In structured text, we can use the `CASE` command.
(16) In sequential function charts, `N` means that as soon as the step is ended, the action is stopped.
(17) In sequential function charts, `R` means that an action is delayed by the given time.
(18) In sequential function charts, it is possible to determine the validity of the resulting chart.
(19) Ladder logic is a great idea to create complex, high-level PLC programmes.
(20) Instruction list is designed for simple, optimised PLC programmes.

7.10.2 *Short Questions*

These questions should be solved using only a simple, nonprogrammable, non-graphical calculator combined with pen and paper.

(21) Is the sequential function chart in Fig. 7.25 valid?
(22) Give the sequential function charts for the controlling the following systems:

a. **Stirring process**: Fig. 7.26 shows the P&ID diagram of a stirred tank B with flow controller `US2`. The binary input signals to `US2` are the high-level alarm `L`, the start signal `S` and the end signal `E`. The binary output signals from `US2` are `V1` for switching the input valve, `V2` for switching the drain valve, and `R` for switching the stirrer motor. The following sequence is required: after the start signal reaches 1 (`S`=1), a liquid is to be dosed by opening the input valve (with `V1`=1) and turning on the stirrer (`R`=1) until the level `L` reaches its upper limit (`L`=1). Then, the valves are closed and the stirrer should alternate between being switched on for 1 min and switched off for 1 min. This

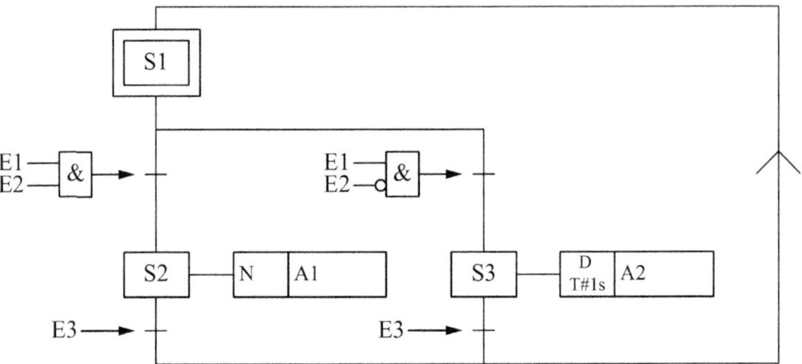

Fig. 7.25 Validity of the sequential function chart

switching on and off should, regardless of the current cycle, be ended immediately when the end button is pressed (`E` = 1). Then, the drain valve should be opened for 5 min. Then, everything should be switched into the idle state, that is, `V1` = `V2` = `R` = 0. When `S` = 1, the process is started from the beginning.

b. **Washing Machine**: The control for a simple washing machine, shown in Fig. 7.27 whose variables are defined in Table 7.14, should implement the following sequence. After the input signal `START` has a value of `1`, valve `V1` should be opened until the binary level sensor `L1` reports that the desired water level has been reached. Then, the `MOTOR` of the washing drum should be switched on alternately for a period of time `T_on` and switched off for a period of time `T_off`. During this alternating switching on and off, the electrical `HEATING` is first switched on until the water temperature has reached a certain value, which is reported by the binary sensor `T2`. Then, the process should wait a certain minimum waiting time `T_wait` before any further heating occurs. Once the motor has stopped running, the alternative

Fig. 7.26 Stirring process

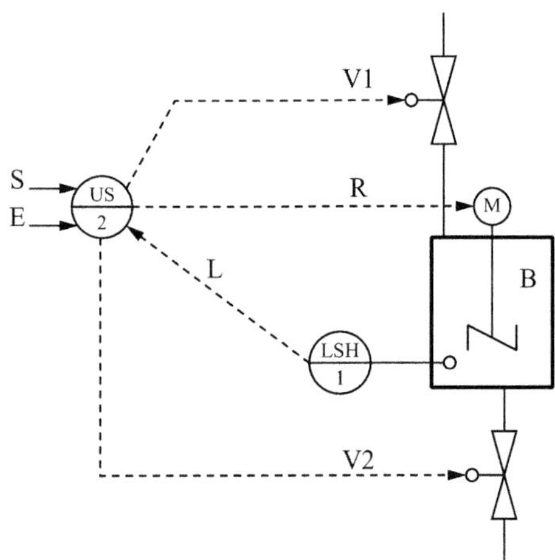

switching of the heating is also stopped. Then, the `PUMP` is switched on and valve `V2` is opened at the same time. Once the level sensor `L0` reports an empty drum, the initial state is reached, and the process can be repeated.

(23) In the examples below, various Boolean functions of the form $Q = f(X, Y, Z)$ are given. For each, please provide (a) the function in its sum-of-products and product-of-sums forms, (b) the reduced function found using a Karnaugh diagram and (c) the PLC programme using instruction list, ladder logic, and function-block language for the minimised function.

Fig. 7.27 Washing machine

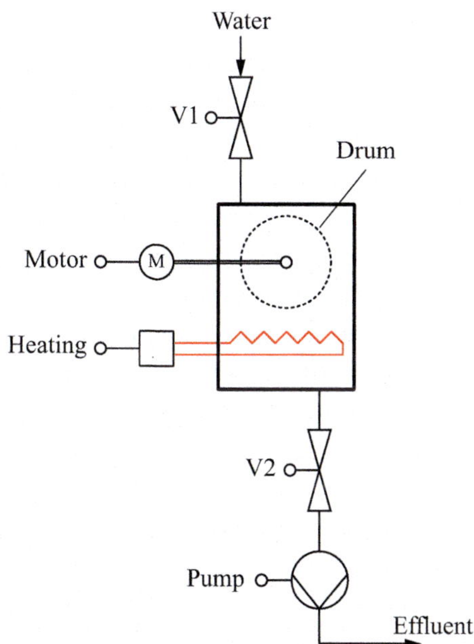

Table 7.14 I/O for the washing machine

I/O	Name	Description for a Boolean Value 1
Output	V1	Valve V1 is open
Output	V2	Valve V2 is open
Output	HEATING	Heating is on
Output	MOTOR	Motor is on
Output	PUMP	Pump is on
Input	START	Start button has been pressed
Input	L0	Level $L \leq L_{min}$
Input	L1	Level $L \geq L_{max}$
Input	T2	Temperature $T \geq T_{desired}$

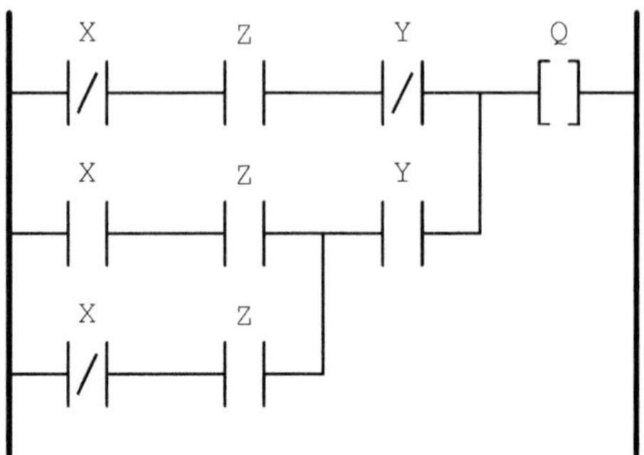

Fig. 7.28 Ladder logic

a. The function shown using ladder logic in Fig. 7.28.
b. The function that implements a majority decision rule for a 2-out-of-3 redundancy system, that is, the output Q is 1 if and only if at least two of the inputs are 1.
c. The function shown by the truth table in Table 7.15.
d. The function shown by the truth table in Table 7.16
e. The function $Q = f(\text{B, G, M})$ is described as follows:

i. Three sensors in a sorting system measure the properties of parts and deliver the following binary signals to the inputs of a PLC:

1. **B**, where B = 1 means a hole is present,
2. **G**, where G = 1 means green paint is present, and
3. **M**, where M = 1 means a metallic material is present.

ii. Parts produced without errors have the following properties:

1. Either metallic and green (with or without a hole)
2. Nonmetallic, not green, and with a hole.

Table 7.15 Truth table I

X	0	0	0	0	1	1	1	1
Y	0	0	1	1	0	0	1	1
Z	0	1	0	1	0	1	0	1
Q	0	1	1	0	1	1	1	1

Table 7.16 Truth table II

X	0	0	0	0	1	1	1	1
Y	0	0	1	1	0	0	1	1
Z	0	1	0	1	0	1	0	1
Q	0	0	1	1	1	0	1	1

iii. The output Q of the PLC should be 1 when the part is to be rejected, i.e. if it does not have the above properties of a faultlessly produced part.

7.10.3 Computational Exercises

The following problems should be solved with the help of a computer and appropriate software packages, such as CodeSys.

(24) Using CodeSys, write the PLC programme to control a filling station.

Chapter 8
Safety in the Automation Industry

As automation becomes more widespread, there is a corresponding need to consider the safety of those who come in contact with the automated systems. Furthermore, automated systems need to take into consideration safety constraints themselves when they implement various actions. For these reasons, it is helpful to briefly review the different safety regulations. Obviously, it is always your engineering obligation to make sure that you have checked the most recent and most relevant regulations and laws applying to your specific situation.

In general, safety can be divided into two parts: **physical** and **digital**. Physical safety considers the steps and regulations required to ensure that the plant and its surroundings are safe for the workers, visitors, and the environment. Environmental protection considers not only hazardous substances, but also the emission of electromagnetic radiation or noise. Digital safety considers the steps required to ensure that the communication networks and the associated devices cannot be hacked and used to create an unsafe physical situation. Furthermore, digital safety must ensure that no sensitive information can be stolen or used in an inappropriate manner. This aspect is very important given the increasing legislative interest in making sure that confidential personal information remains confidential.

8.1 Safety of the Physical System

Before delving into the specifics of physical safety, it is useful to consider some basic principles. A **hazard** is a source of danger, while a **risk** is the degree or extent of peril that a hazard could create. Hazards arise from such things as moving objects, stored energy, and explosions. Essentially, hazards primarily arise from the release of latent (or hidden stored) energy. These include:

Y. A. W. Shardt, *Automation Engineering*,
https://doi.org/10.1007/978-3-031-92394-4_8

(1) **Kinetic Energy**, which is the energy resulting from moving objects, can present hazards in any rotating equipment, such as pumps and turbines, vehicles, and conveyor belts.
(2) **Potential Energy**, which is the energy resulting when an object unexpectedly falls. Such releases of energy are common in structural failures, such as building collapses.
(3) **Work**, which is energy stored in springs, electrical circuits, and other devices. When released, this can lead to catastrophic failure and impressive damage. A common electrical example is a short circuit.
(4) **Heat**, which is energy crossing a boundary. Heat exchange is always based on a temperature difference between two points in space. Since heat always influences the environment, touching hot surfaces, for example, can lead to burns. The same situation applies to touching very cold surfaces, which can also lead to burns or frostbite.

In order to increase the safety of a physical system, it can be useful to consider the following steps when designing the overall system:

(1) **Minimise**: Avoid using (or use as little as possible) any hazardous substances.
(2) **Substitute**: Replace hazardous substances by ones that are less hazardous but with similar properties.
(3) **Moderate**: Replace extreme operating conditions by ones that are less severe.
(4) **Simplify**: Create designs or processes that are less complex and simpler.
(5) **Isolate**: Create situations where the effect of a hazard has minimal effect on the overall system, for example, physically separating the office space from the production area.

The above ideas can be framed within the **inherently safer predesign** (ISPD) approach, which consists of the following 4 steps:

(1) **Identify**: Determine what the hazards are and how they can affect the process.
(2) **Eradicate**: Remove the effects of as many of the hazards as possible, for example, by designing **fail-safe** systems, that is, should a system encounter a failure it ends up in an inherently safe mode that cannot cause further problems. A common example of this is the design of control valves to fail open or closed depending on the process conditions, for example, a control valve for a critical cooling jacket should fail open so that the system remains cooled even if the valve has failed.
(3) **Minimise, simplify, and moderate**: If a hazard cannot be eradicated, then try and limit its effects on the system.
(4) **Isolate**: Create structures that will limit the amount of damage that can occur, for example, by segregating hazardous operations.

Another approach to this situation is to perform a **hazard-and-operability study** (**HAZOP**) to determine where the given hazards lie and how best to mitigate them. The HAZOP procedure consists of the following four steps: determine, using a systematic search, possible hazards; find the causes for the hazards; estimate the effects of the hazards on the system; and suggest appropriate countermeasures. A HAZOP requires that a detailed plan of the process be available. Thus, such analysis is often performed at a later stage of design after the overall process design has already been completed. This implies that, due to the high cost or structural changes required, it could be too late to make substantive changes at this point. Thus, HAZOPs are often used to provide an understanding of the hazards present and their risk.

Finally, it can be mentioned that when designing safety systems, it is useful to consider **redundancy**, that is, having more than one channel or way to achieve a given task. As well, this implies that different channels should be used for different tasks (control, alarm, and monitoring), so that there is less risk of catastrophic failure. Thus, for example, alarms and monitoring should not occur on the same channel. If it should fail, alarm signals can no longer be transmitted in addition to the less important monitoring signals. There are two main ways to implement redundancy. In the first approach, often called **homogeneous redundancy**, the same task is implemented by multiple (identical or similar) devices, for example, the temperature is measured using three temperature sensors at the same point. However, this approach carries the risk that systematic errors in the sensors or devices are propagated into the system. In the second approach, often called **heterogeneousredundancy** or **redundancy by diversity**, the same task is implemented using different devices, for example, three separate path-planning computer systems that use different computational algorithms could be implemented and the decision taken by majority vote, that is, if two or more algorithms give the same result, it is followed. Selecting the appropriate approach depends on the requirements and standards of the process.

In order to properly understand and implement a safety system, it is necessary to quantify the risk in some manner. One approach is to define the risk R as

$$R = CD \tag{8.1}$$

where C is the impact of the hazard and D is the frequency of the hazard. The frequency of the hazard is defined as

$$D = FPW \tag{8.2}$$

where F is the occurrence frequency, P is the probability that the hazard cannot be mitigated, and W is the probability that without any safety system in place an undesired state will be reached. Since it makes little sense to define the risk with an exact value, it is common to instead associate a given level to the individual variables and look at the combination of levels.

For C, the following levels are defined:

- **C1**: Minor injuries
- **C2**: Major or permanent injuries of one or more people or a single death
- **C3**: Up to five deaths
- **C4**: More than five deaths.

For *F*, two levels are defined:

- **F1**: During the course of the day, people are in the danger zone less than or equal to 10% of the time.
- **F2**: During the course of the day, people are in the danger zone more than 10% of the time.

For *P*, two levels are defined:

- **P1**: It is possible to mitigate the hazard (one should provide the appropriate steps to do so).
- **P2**: It is not possible to mitigate the hazard.

For *W*, three levels are defined:

- **W1**: The undesired state will occur less than once every ten years.
- **W2**: The undesired state will occur less than once per year.
- **W3**: The undesired state will occur more often than once per year.

The above levels are then combined to give an overall **safety integrity level** (SIL). There are four safety integrity levels labelled SIL1 to SIL4, where the larger number reflects a more hazardous level. The relationship between the different parameters is shown in Fig. 8.1.

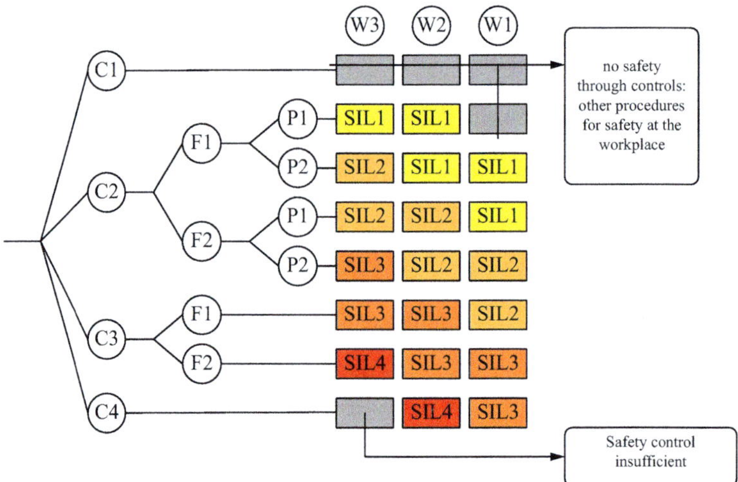

Fig. 8.1 Relationship between the parameters and the safety-integrity levels

8.2 Safety Regulations

Safety regulations in Germany (and most of Europe and even the world) come in many different flavours and legal force. In Germany, standards are developed by the German Institute for Standardisation (DE: Deutsches Institut für Normung, commonly abbreviated as DIN). DIN publishes standards not only for safety but for all sorts of other aspects, such as formatting of letters, transliteration, and engineering. All standards developed or accepted by DIN are prefixed by the letters *DIN* followed by a number. Additional letters can be added to show the general applicability of the standards. Using *DIN EN* implies that we are dealing with a German edition of a European standard, while *DIN ISO* implies that we are dealing with a German edition of an international standard developed by the International Standards Organisation (ISO). Obviously, it is possible to have a *DIN EN ISO* label which would denote a German edition of a European and international standard. In general, when there are multiple prefixes, this implies that we are dealing with the same standard under different guises, for example, a DIN ISO #### standard would be the same as the corresponding ISO ####. Examples of different DIN standards are:

(1) DIN 31635 which standardises the transliteration of the Arabic script used to write Arabic, Ottoman Turkish, Iranian, Kurdish, Urdu, and Pasto.
(2) DIN EN ISO 216 which standardises writing paper and certain classes of printed matter, trimmed sizes, A and B series, and indication of machine direction.
(3) DIN ISO 509 which covers the production of technical drawings, relief grooves, and types and dimensions of them.
(4) DIN EN 772-7 which covers clay masonry units, namely the determination of water absorption of clay masonry damp proof course units by boiling water.

Other countries use similar systems, for example, Austria prefixes its codes with ÖNORM while the United States of America uses the ANSI system.

Safety regulations in Germany (and most of Europe) for machines can be split into three types:

- **Type A Regulations**: These regulations cover the basic safety concepts, design principles, and general aspects of machines. An example of a Type A regulation is the EN ISO 12100 standard that provides the general principles of design.
- **Type B Regulations**: These regulations are baseline safety standards and requirements for protective equipment. There are two subtypes: **B1-Regulations** that cover specific safety aspects and **B2-Regulations** that cover the standards for protective equipment required for machines. An example of a Type B1 regulation is the EN ISO 13855 standard that provides the guidelines for the arrangement of protective devices, while a Type B2 regulation would be the EN 953 standard that considers fixed guards for machines.

- **Type C Regulations**: These regulations are the specific safety standards for a particular machine or group of machines. An example of a Type C regulation would be the EN 693 standard that provides the standards for hydraulic presses.

8.3 Digital Safety

In today's world, where many components of a plant are now linked digitally using networks, it is imperative that the digital system be secured. The main topic in digital safety focuses on preventing intruders from entering the digital system and making unauthorised changes to the system, that is, digital safety must prevent **hacking** from occurring. Hacking can expose a company to various threats including loss of proprietary information, damage to the process by illegal changing of process operating conditions, and legal issues due to distribution of confidential personal information.

Digital security can be achieved by creating strong passwords, changing the passwords on a regular basis, and checking for any intruders. Strong antivirus software should also be implemented. Multiple interacting networks should be considered to provide additional levels of security. Finally, the workers should be educated in proper digital safety as the digital system is only as strong as its weakest link. People placing unknown USB-drivers into work computers can easily introduce unwanted viruses into the company network.

8.4 Further Reading

The following are additional resources on the topics mentioned in this chapter:

(1) G. D. Ulrich and P. T. Vasudevan (2004). *Chemical Engineering Process Design and Economics: A Practical Guide* (Second Edition), Durham, New Hampshire, USA: Process Publishing.

(2) Don W. Green (editor) (2018). *Perry's Chemical Engineers' Handbook.* (85th Edition), New York, New York, USA: McGraw-Hill.

8.5 Chapter Problems

Problems at the end of the chapter consist of two different types: (a) Basic Concepts (True/False), which seek to test the reader's comprehension of the key concepts in the chapter; and (b) Short Exercises, which seek to test the reader's ability to compute the required parameters for a simple data set using simple or no technological aids.

8.5.1 Basic Concepts

Determine if the following statements are true or false and state why this is the case.

(1) Process safety is never a topic of concern.
(2) A hazard is something that has a high likelihood of causing harm.
(3) A short circuit is an example of a hazard caused by the release of stored energy or work.
(4) Collapsing buildings are an example of hazards posed by kinetic energy.
(5) Hazards that cannot be eliminated should be isolated.
(6) Designing complex processes is a good risk-mitigation strategy.
(7) The fail-safe principle states that a failing system should always end up in a safe mode.
(8) Using the fail-safe principle, valves should always fail closed.
(9) A HAZOP seeks to identify the risks and how best to mitigate them.
(10) Redundancy means that a single temperature sensor is used for process monitoring and process control.
(11) Selecting a course of action based on two-out-of-three voting is an example of redundancy by diversity.
(12) A risk with an occurrence frequency greater than 10% is always an SIL 4 risk.
(13) A risk with a hazard-impact level of C3 implies that it can cause only minor injuries.
(14) A risk with a mitigation level of P2 implies that it cannot be mitigated.
(15) A risk with a *W* level of W3 implies that the undesired state will occur more often than once per year.
(16) In Germany, Type A regulations cover the detailed requirements for the safety of a specific machine.
(17) An example of a Type B2 regulation would be the design of protective guards for machines.
(18) An example of a Type C regulation would be the standards for the design of chemical reactors.
(19) Using simple passwords and minimal digital security is a good idea for a critical chemical process.

8.5.2 Short Questions

These questions should be solved using only a simple, nonprogrammable, nongraphical calculator combined with pen and paper.

(20) Determine the safety integrity levels for the following risks:

a. A risk evaluated as belonging to C2, F1, W2, and P2.
b. A risk evaluated as belonging to C3, F2, P1, and W1.

(21) Determine some of the safety issues involved with the production of the following chemicals. Based on the principles of safety considered here, would you recommend that these chemicals be produced by a new start-up that has no chemical engineering background.

a. Acrylonitrile (CH_2CHCN), which is produced from propene, air, and ammonia.
b. Propene (C_3H_6), which is produced from ethene and 2-butene.

(22) Perform a HAZOP on the compressor unit shown in Fig. 8.2.

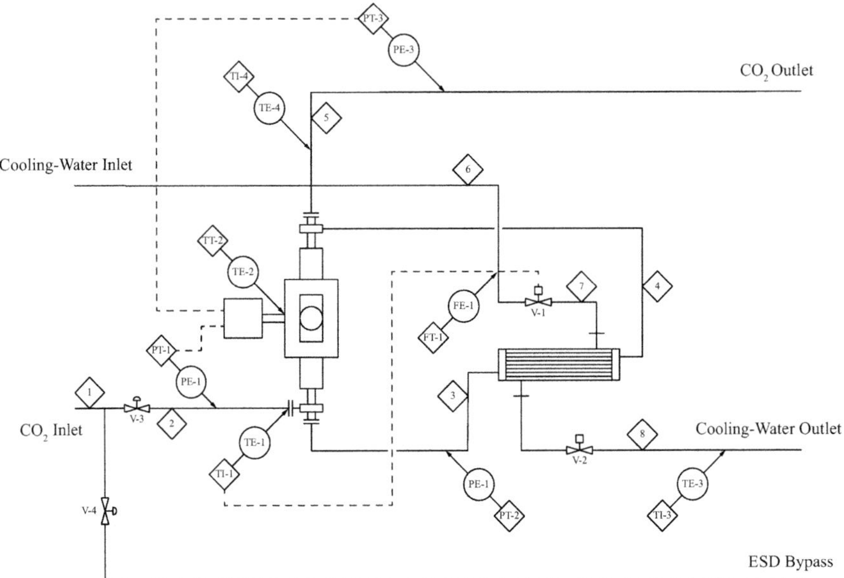

Fig. 8.2 P&ID for a compressor unit

Appendix

Partial Fractioning

When solving equations in the frequency domain and it is desired to convert them back into the time domain, it may be necessary to perform **partial fractioning** to obtain a solution. Although there exist many different approaches, the following is one simple method that will allow the final result to be obtained easily. Consider a rational function of the form:

$$\frac{N(s)}{D(s)} = \frac{N(s)}{\prod_{i=1}^{n_l} (\alpha_i s + \beta_i)^{n_i} \prod_{j=1}^{n_q} \left(\alpha_j s^2 + \beta_j s + \gamma_j\right)^{n_j}} \tag{A.1}$$

where n_l represents the number of distinct linear terms, n_q the number of distinct irreducible quadratics (those that have imaginary roots as their solution), and $\alpha, \beta,$ andγ are known constants. Let n be the overall order of the system. In order to perform partial fractioning, write the following fraction depending on the form of the root:

(1) For each linear term $(\alpha s + \beta)^n$, put the term, $\sum_{k=1}^{n} \frac{B_k}{(\alpha s+\beta)^k}$.

(2) For each irreducible quadratic term $\left(\alpha s^2 + \beta s + \gamma\right)^m$, put the term, $\sum_{k=1}^{m} \frac{A_k+B_k s}{\left(\alpha s^2+\beta s+\gamma\right)^k}$.

Once the form of the partial fractioning solution has been obtained, it is necessary to solve for the unknown parameters. First, cross-multiply, so that the denominators are the same. Then, it is necessary to solve for the unknown parameters by equating the unknown side with the known *N*(*s*). The easiest way to solve this is using the following approach:

Y. Shardt, *Automation Engineering*, https://doi.org/10.1007/978-3-031-92394-4

(1) For each linear term, set $s = -\beta/\alpha$ to obtain B_n of the linear terms. This will reduce the equation to the form $B_n \prod_{\substack{i=1 \\ i\neq n}}^{n_l} (\alpha_i s + \beta_i)^{n_i} \prod_{j=1}^{n_q} \left(\alpha_j s^2 + \beta_j s + \gamma_j\right)^{n_j} = N(s)$ evaluated at the given root.
(2) For each quadratic term, set s equal to the imaginary roots. This will also reduce the equation to a simpler form and allow for A_n and B_n to be solved.

For the remaining terms, create a system of equations by selecting different values of s and evaluating the known values, so that the remaining unknowns can be solved. You will need $n - n_l - 2n_q$ equations in order to find the remaining $n - n_l - 2n_q$ terms.

Example A.1: Partial Fractioning Consider the following fraction

$$\frac{3s+1}{(s+2)(s+1)^2\left(s^2+1\right)} \tag{A.2}$$

for which we wish to determine the partial fraction form.

Solution

First, we need to write the general form, that is, for each of the terms in the denominator, we will write the corresponding partial fraction using the rules above. This will give

$$\frac{A}{s+2} + \frac{B}{s+1} + \frac{C}{(s+1)^2} + \frac{Ds+E}{s^2+1} \tag{A.3}$$

It can be noted that for the first term $s + 2$, as well as the last term $s^2 + 1$, there will only be a single component, since its exponent is one. For the middle term, $(s + 1)^2$, there will be two terms, since the exponent is two. For the linear terms, we set a simple constant term in the numerator, while for the quadratic term, we include a linear term in the numerator.

Next, we need to determine the values of the constants in the numerator. Before doing this, let us cross-multiply and determine the general form of the numerator

$$\begin{aligned} &A(s+1)^2\left(s^2+1\right) + B(s+2)(s+1)\left(s^2+1\right) \\ &+ C(s+2)\left(s^2+1\right) + (Ds+E)(s+2)(s+1)^2 = 3s+1 \end{aligned} \tag{A.4}$$

As was previously mentioned, we can see that setting $s = -1$ or -2 will cancel every term but one, allowing us to effortless compute that term. This gives for $s = -1$

$$C(-1+2)\big((-1)^2+1\big) = 3(-1)+1$$
$$C = \frac{-2}{2} = -1 \tag{A.5}$$

Similarly, for $s = -2$, we get

$$A(-2+1)^2\big((-2)^2+1\big) = 3(-2)+1$$
$$A = \frac{-5}{5} = -1 \tag{A.6}$$

For the quadratic term, we can set $s = \pm j$, which will give us a linear system of equations in two unknown (D and E) that we can then solve to obtain the value of D and E. This gives

$$\begin{cases} (Dj+E)(j+2)(j+1)^2 = 3j+1 \\ (-Dj+E)(-j+2)(-j+1)^2 = -3j+1 \end{cases} \tag{A.7}$$

Placing all the constant terms on the right gives

$$\begin{cases} Dj+E = 0.5-0.5j \\ -Dj+E = 0.5+0.5j \end{cases} \tag{A.8}$$

Adding the two equations together gives that $E = (0.5+0.5)/2 = 0.5$. From this, it follows that $D = -0.5$. It can be noted that these two equations will always be complex conjugates of each other and one can use this fact to solve them without necessarily computing both components.

The remaining term B can be found by selecting an arbitrary value of s (that has not already been used) and solving Eq. (A.4). The known constants are inserted as required. Setting $s=0$, we get

$$-1(1)^2(1) + B(2)(1)(1) - 1(2)(1) + (0.5)(2)(1)^2 = 1$$
$$B = \frac{3}{2} = 1.5 \tag{A.9}$$

Obviously, we have the correct answer if the numerator given by Eq. (A.4) holds. This can be used to check the solution obtained.

As per instruction book end references have been moved to chapters and remaining uncited references are retained in this chapter under the heading Bibliography. Please check and provide the citations or delete these references.I am not sure what this statement is supposed to mean, since none of the references that have been cited have been moved to the end of the chapters. Personally, I would prefer maintaining the bibliography as is. All of the references are cited in the chapters (either directly) or in the further reading section. Therefore, I need more information before I can proceed to answer this query.

Bibliography

1. Åström KJ, Hägglund T (1995) PID controllers: theory, design, and tuning, 2nd edn. Instrument Society of America, Research Triangle Park, North Carolina, USA
2. Carroll J, Long D (1989) Theory of finite automata with an introduction to formal languages. Prentice-Hall, Englewood Cliffs, New Jersey, USA
3. Cassandras CG, Lafortune S (2008) Introduction to discrete event systems, 2nd edn. Springer, New York, USA
4. Doetsch G (1974) Introduction to the theory and application of the Laplace transform. Springer, Berlin, Germany
5. El Attar R (2004) Lecture notes on Z-transform. Lulu Press, Morrisville, North Carolina, USA
6. Green DW (ed) (2018) Perry's chemical engineers' handbook, 85th edn. McGraw-Hill, New York, USA
7. Halmos P, Givant S (2009) Introduction to Boolean algebras. Springer, New York, USA
8. John K-H, Tiegelkamp K (2009) IEC 61131–3: programming industrial automation systems: concepts and programming languages, requirements for programming systems, decision-making aids, 2nd edn. Springer, Berlin, Germany
9. Kreyszig E (2011) Advanced engineering mathematics, 10th edn. Wiley, New York, USA
10. Liu L, Tian S, Xue D, Zhang T, Chen Y (2019) Industrial feedforward control technology: a review. J Intell Manuf 30:2819–2833
11. Lundberg KH, Miller HR, Trumper DL (2007) Initial conditions, generalizerd functions, and the Laplace transform: troubles at the origin. IEEE Control Syst Mag 27(1):22–35. https://doi.org/10.1109/MCS.2007.284506
12. Ogata K (1995) Discrete-time control systems, 2nd edn. Prentice-Hall, Upper Saddle River, New Jersey, USA
13. Rawlings JB, Mayne DQ, Diehl MM (2017) Model predictive control: theory, computation, and design, 2nd edn. Nob Hill Publishing, Madison, United States of America
14. Seborg DE, Edgar TF, Mellichamp DA, Doyle FJ (2011) Process dynamics and control, 3rd edn. Wiley & Sons Inc., Hoboken, New Jersey, United States of America
15. Shardt YA (2012) Data quality assessment for closed-loop system identification and forecasting with application to soft sensors. Edmonton, Alberta, Canada, University of Alberta. http://hdl.handle.net/10402/era.29018
16. Shardt YAW, Zhao Y, Qi F, Lee KH, Yu X, Huang B, Shah S (2012) Determining the state of a process control system: current trends and future challenges. Canadian J Chem Eng 217–245. https://doi.org/10.1002/cjce.20653
17. Shoukat Choudhury MA, Thornhill NF, Shah SL (2005) Modelling valve stiction. Control Eng Pract 641–658
18. Skogestad S (2023) Advanced control using decomposition and simple elements. Annu Rev Control 56:100903. https://doi.org/10.1016/j.arcontrol.2023.100903

Y. Shardt, *Automation Engineering*, https://doi.org/10.1007/978-3-031-92394-4

19. Sontag ED (1998) Mathematical control theory: deterministic finite dimensional systems. Springer, New York, USA
20. Toghraei M (2019) Piping and instrumentation diagram development. Wiley, Hoboken, New Jersey, USA
21. Ulrich GD, Vasudevan PT (2004) Chemical engineering process design and economics: a practical guide, 2nd edn. Process Publishing, Durham, New Hampshire, USA
22. Visioli A (2006) Practical PID control. Springer, London, UK

Index

Y. Shardt, *Automation Engineering*, https://doi.org/10.1007/978-3-031-92394-4

The manufacturer's authorised representative in the EU is Springer Nature Customer Service Centre GmbH, Europaplatz 3, 69115 Heidelberg, Germany. If you have any concerns regarding our products, please contact ProductSafety@springernature.com

Printed and bound by CPI Group (UK) Ltd, Croydon, CR0 4YY
22/07/2026
02174115-0001